ONLINE CHINESE NATIONALISM AND CHINA'S BILATERAL RELATIONS

Challenges Facing Chinese Political Development

Series Editor: Sujian Guo, PhD
San Francisco State University

In an attempt to reflect the rapidly changing political environment of the People's Republic of China, editor Sujian Guo has assembled a book series to present specialized areas of research in current Chinese political studies. Incorporating theoretical, empirical, and policy research on contemporary Chinese politics both domestically and internationally, this series contemplates the Chinese past, present, and future by utilizing interdisciplinary perspectives to approach issues related to Chinese politics, economy, culture, social development, reform, the military, legal system, and foreign relations. Aimed at bringing a greater understanding of the current Chinese political climate to Western audiences, this series is focused on the emerging voices of Chinese scholars and their perspectives on the ever-changing Chinese diaspora.

Recent titles in the series are:

Challenges Facing Chinese Political Development, edited by Sujian Guo and Baogang Guo

New Dimensions of Chinese Foreign Policy, edited by Sujian Guo and Shiping Hua

The Dragon's Hidden Wings: How China Rises with Its Soft Power, by Sheng Ding

"Harmonious World" and China's New Foreign Policy, by Sujian Guo and Jean-Marc F. Blanchard

China in Search of a Harmonious Society, edited by Sujian Guo and Baogang Guo

Greater China in an Era of Globalization, edited by Sujian Guo and Baogang Guo

Toward Better Governance in China, edited by Baogang Guo and Dennis Hickey

Dynamics of Local Governance in China during the Reform Era, edited by Tse-Kang Leng and Yun-han Chu

Dancing with the Dragon: China's Emergence in the Developing World, edited by Dennis Hickey and Baogang Guo

Online Chinese Nationalism and China's Bilateral Relations, edited by Simon Shen and Shaun Breslin

ONLINE CHINESE NATIONALISM AND CHINA'S BILATERAL RELATIONS

Edited by
Simon Shen and Shaun Breslin

LEXINGTON BOOKS
A division of
ROWMAN & LITTLEFIELD PUBLISHERS, INC.
Lanham • Boulder • New York • Toronto • Plymouth, UK

Published by Lexington Books
A division of Rowman & Littlefield Publishers, Inc.
A wholly owned subsidiary of The Rowman & Littlefield Publishing Group, Inc.
4501 Forbes Boulevard, Suite 200, Lanham, Maryland 20706
www.lexingtonbooks.com

Estover Road, Plymouth PL6 7PY, United Kingdom

British Library Cataloguing in Publication Information Available

Library of Congress Cataloging-in-Publication Data

Online Chinese nationalism and China's bilateral relations / edited by Simon Shen and Shaun Breslin.
p. cm.
Includes index.
ISBN 978-0-7391-3247-0 (cloth : alk. paper) — ISBN 978-0-7391-3249-4 (electronic)
1. Internet—Political aspects—China. 2. Nationalism—China. 3. China—Foreign relations. 4. Civil society—China. 5. Information policy—China. I. Shen, Simon. II. Breslin, Shaun.
HN740.Z9I56788 2010
303.48'30951—dc22 2009051171

™ The paper used in this publication meets the minimum requirements of American National Standard for Information Sciences—Permanence of Paper for Printed Library Materials, ANSI/NISO Z39.48-1992.

Printed in the United States of America

Contents

Figures

Tables

Acknowledgments

The editors of the book acknowledge the International Relations Research Programme of the Hong Kong Institute of Asia-Pacific Studies, the Chinese University of Hong Kong, and the Association of Chinese Political Studies, as well as the International Relations Research Association of the Roundtable Social Science Network of Hong Kong for co-organizing an international academic conference on "China in the Era of Globalization" in July 2008, which resulted in the formation of two special panels dedicated to the book's topic of online Chinese nationalism. Several contributors who did not attend the conference are equally significant to the book, as their contributions have greatly enriched the content of the collection and made the book really multidimensional. Special thanks are also due to research assistant Mr. Wilson Chan, whose meticulous communication with the several contributors and copyeditors deserves full credit. Finally, funding support from the South China Programme of the Chinese University of Hong Kong is fully appreciated.

I

INTRODUCTION

1

When China Plugged In: Structural Origins of Online Chinese Nationalism

Simon Shen and Shaun Breslin

> In the remote but beautiful *Malegebi*, there is a group of grass-mud horses (*caonima*).[1]

Introduction

On 12 June 2009, the Chinese government suddenly announced a new and immediate standard for computers manufactured and sold in the mainland: a mandatory Internet filtering software system, *luba-huaji huhang* (the Green Dam), was to be installed from 1 July. According to a Chinese Ministry of Industry and Information Technology (MIIT) spokesperson, the primary objective for the mandatory installation of the software was to screen out inappropriate material and pornographic websites, so as to "create a green and healthy Internet environment" within the virtual community.[2] The announcement of the measure was quickly followed by considerable criticism, both from within China and without. For instance, U.S. Secretary of Commerce Gary Locke and Trade Representative Ron Kirk sent a joint letter to MIIT and China's Ministry of Commerce, saying that the scope and the extent of the filtering activities raised "fundamental questions regarding regulatory transparency."[3] In China, various "anonymous" netizens criticized the government for its de facto infringement of the freedom of expression and right to receive information, and decided to launch a series of attacks on the Chinese Communist Party–led Internet censorship.[4] As a result of the fervent opposition to the scheme, Beijing finally postponed the implementation of the measure just one day before it was due to be introduced. Nonetheless, the controversy this proposal caused has fully exposed the friction between the official agenda and the online civil society.

Since the Chinese were officially plugged into the virtual community in 1994, the use of the Internet in the country has developed at an incredible rate. By the end of 2008, there were approximately 298 million netizens in China, a number which surpasses that of the United States and ranks China the highest user in the world.[5] The rapid development of the online Chinese community has not only boosted the information flow among citizens across the territory, but has also created a new form of social interaction between the state, the media, various professionals and intellectuals, as well as China's ordinary citizens. Although the subject of this book is online Chinese nationalism, which to a certain extent is seen as a pro-regime phenomenon, the emergence of an online civil society in China intrinsically provides some form of supervision of state power—perhaps even a check on it. The fact that the party-state has made use of this social interaction while at the same time remaining worried about the negative impact of the same netizens is a fundamental characteristic of the nature of the relationship between the state and the Internet community.

The Nascent Public Sphere in Authoritarian China

As Andrew Chadwick comments in *Internet Politics*, "the issue is no longer whether politics is online, but in what form and with what consequences?"[6] This has become a globalized phenomenon applicable to almost all nations in the world. The launching of Web 2.0, which has enabled the evolution from static web pages to multimedia and interactive websites, has also furthered politics on the Internet. In their *Handbook of Internet Politics*, Andrew Chadwick and Philip Howard list the new features of Internet politics to be found in the developed world, modified versions of which are certainly also applicable to China.[7]

First of all, the Internet serves as a platform for political discourse. The value of Internet portals like Google or Yahoo! depends on their capacity to create the interface of their distributed networks, basically an infrastructure providing a huge information database and networked crawled pages. In other words, the Internet provides a platform for netizens to flexibly adapt to sudden political changes. As a result, political campaigns are now being mobilized online. For instance, in January 2007, U.S. politician John Edwards used YouTube to announce his candidacy for the Democratic presidential nomination; the popularity of Barack Obama has also been hugely boosted by online mobilization, and much use has been made of YouTube, Facebook, and blogs for fundraising and campaigning. In China, the party-state has also attempted to dissimilate information online in a top-down manner. However, the strong control of the party-state results in more pro-government messages being found online, making it difficult for researchers to differentiate state-sponsored online messages from voluntarily patriotic ones.

Another feature of Internet politics 2.0 relevant to our topic is the idea of collective intelligence. The main thought here is that by gathering collective information from a critical mass, a widespread network of amateurs can potentially outperform the official discourse prepared by the authorities. One example in East Asia is the culture of *Kuso* (or in Chinese, *e gao*) in Japan, Taiwan, Hong Kong, and mainland China. Originating in Japan, *Kuso* is defined as "making fun of everything and playing practical jokes."[8] For instance, messages given by pro-Beijing politicians in Hong Kong are often satirized by netizens by means of *Kuso*. Another obvious example that exhibits the power of collective intelligence is the creation of bottom-up interactive encyclopedias such as Wikipedia and its Chinese counterpart, *baidu*. Supporters of different political candidates, regimes, causes, movements, and so on have engaged themselves in "edit wars" in these encyclopedias. The blogsphere also allows fellow bloggers to bypass traditional printed media to express themselves to the public. In short, Web 2.0 further empowers ordinary citizens to participate in public affairs. This is of considerable significance in authoritarian regimes like China, as citizen access to high politics is normally largely blocked. In fact, the Internet has become the only platform for alternative messages to be circulated and reconstructed in China in a bottom-up manner, albeit that they are often disguised in patriotic colors in order to get through the official censorship. This feature is one of the key differences between the Chinese cyberspace and that of the West.

Then what are the overriding features of China's online civil society? When Guobin Yang gave his preliminary assessment of the relationship between Chinese civil society and the Internet, he characterized the virtual community in China by the Habermasian concept of a "public sphere."[9] This "public sphere," as suggested by Jürgen Habermas, has several characteristics that offer: (1) a platform for rational debate among autonomous and rational individuals; (2) spaces where the public are free to engage and disengage themselves from the debate; and (3) a medium of communication, such as newspapers and books, that are necessary for deliberation.[10] While Habermas describes the public space as being similar to the café salon of the French Revolution period, the creation of the Internet space in China encourages the involvement of laymen making it a truly public sphere. Identified by Yongnian Zheng and Xiaoling Zhang, one key issue of cyberpolitics in China is whether civil society, which is still a nascent entity in an undemocratic China, could be empowered in its relation to the state, or could become a "digital civil society."[11] Following the traditional definition suggested by Muthiah Alagappa, Guobin Yang defined a "digital civil society" as encompassing "the large numbers of online communities, web-based social networks and loose organizations, as well as the active online presence of offline civic associations."[12]

The year 2003, Yang argues, was a particularly important year for the development of such a digital civil society in China.[13] To start with, there was a great surge in the number of websites and web pages, as shown by the China Internet Network Information Center (CNNIC) statistics.[14] BBS forums, dynamic webpages, and chatrooms started to bloom a decade after China was first plugged in. With the gradual rise in China of Western civic ideas, such as human rights and environmental protection, new forms of civil activism, including rightful resistance (*weiquan*) and environmental protectionism in rural areas, were launched. In addition, the Internet became the platform for Chinese netizens to commemorate various domestic political incidents—for example, the Sun Zhigang, the BMW, and the Liu Yong incidents.[15] While most of the incidents deliberated on the Internet confronted the state machine, the protests of the netizens were quite diversified. One of the most famous examples is that quoted at the beginning of this introduction: the idea of a "grass-mud horse" is used to contrast with another metaphor, that of a "river crab"; the homonym for the former (*caonima*) is one of the most abusive colloquial insults in China while the latter (*hexie*) is a homonym for Hu Jintao's repeated promotion of harmony (as in harmonious society—*hexie shehui*).

Neglected Structural Origins of Online Chinese Nationalism

Of course, the cyber community in China does not just target the party-state. In fact, it is far better known for its nurturing of a fervent dogmatic and patriotic nationalist stance against foreign pressure. The theoretical understanding of Chinese nationalism is also studied here. As Yongnian Zhang has argued, "to build a nation-state from above is the only choice for many late-developing countries like China."[16] An imposed nation-state means the CCP would use all means available to promote the idea of nationalism, especially with the fading of communist ideologies following the open-door policy.[17] Drawing on the empirical evidence in U.S. foreign policy, David Campbell sees foreign policy as always a "boundary-producing performance" which involves the construction of the "Self" and the "Other."[18] A sociological understanding of nationalism may reject such a simple dichotomy; as Alexander

Wendt argues: "if a process is self-organizing, then there is no particular Other to which the Self is related."[19] Nationalism, as one form of collective distinction, is a pure collective identification which, by definition, concerns "a relation of self and other by which the identity of the self is constituted in symbolic markers."[20] Therefore, foreign policy serves as the arena for contest and cooperation between official nationalistic and popular discourse on the question: "What is China/Chinese?" Answers to this question have been transformed and much modified since China "plugged in."

By making foreign policy part of the focus of the book, we acknowledge the idea that international relations represents a consideration of domestic interests, and that giving privilege to some interests over others does not have to proceed from a liberal position. For example, neoclassical realists accept that domestic interests are important in shaping policy, but also argue that the freedom of states to act upon these interests is severely constrained by the nature of the global order; in short, the agency to act is limited by the structure of the international system.[21] On the other hand, like liberals, Marxists also conceive of state action at the international level representing the interests of a subset of society, but focus on class as the cause of societal cleavages and the source of different interests. When searching for contributions to this volume, while we wanted to emphasize the impact of domestic interests on policy formation, we were not prescriptive in terms of which theoretical approach to adopt (beyond a necessary rejection of some extreme realist ontologies). What we sought from each chapter was an identification of the different voices and interests—a taxonomy of nationalisms within the given case study—and the author's own interpretation of the sources of this divergence. Such divergence is exactly what the chapters in this book have individually and collectively uncovered.

The reasons which have led to the possibly exaggerated presence of online nationalists in China, when compared with the developed world, have already been briefly discussed and they will be repeatedly noted in the coming chapters. We also provide a more detailed overview of the collective findings of the chapters and their implications for the study of China in the conclusion. This introduction simply serves to point out that online Chinese nationalism has been understudied by academia. One reason for this seems to be the lack of acknowledgment of the existence of a civil society in China, whether online or not. In its definitional sense, a civil society must have two prerequisites: a pluralistic democracy and a capitalistic economy. Comparing it with the Western-style civil society, Baogang He and Jonathan Unger have coined the civil society in China as a "semi-civil society" or a "hybrid of socialist corporatism and clientelism," the significance of which is not always fully addressed.[22] In addition, methodologically, the state's censorship has created new hurdles for researchers. It is also difficult for us to verify the credibility of online content. As Christopher Hughes observed, the nationalist players do not necessarily have to act upon their nationalist rhetoric.[23]

Indeed, the Chinese authorities have actually been relatively liberal in allowing online debate and discussion that can be very critical of government policy and action. But rather than creating a new political "space" where the promotion of liberalism flourishes, this new "space" has tended to be primarily a forum for the discussion of national interests and the promotion of what might bluntly be called "nationalism." This is not to say that there is a clear and coherent expression of ideology or indeed a clear and coherent set of policy preferences and options, but simply that there is a dominant sentiment that defending national interests in the face of a largely hostile international environment should be the government's main task—and that in some ways the government is not fulfilling this task effectively. Crucially, we argue that there is not a single voice, a single response and a single demand. Rather, there are different responses on a case-by-case level, and the purpose of this book as a whole is to tease these out by the provision of an empirically rich set of case studies. Furthermore, there are also different voices within individual case studies, and this will be a key component of the individual chapters.

These features alone are sufficient to convince ourselves against seeing the online Chinese nationalists as a monolithic group. Instead, no matter how patriotic they appear to be, as Peter Hays Gries concludes, "China's cyber-nationalists, armed with PCs and cell phones, are increasingly contesting party claims to nationalist legitimacy."[24] In a departure from the first wave of the nationalist movement, which was confined to debates by (public) intellectuals of numerous incidents, the "second wave" of online nationalism has emerged from BBS groups such as the Strong Nation Forum (*Qianguo Luntan*). The grassroots version of nationalism, while being equally anti-Western, has at the same time also challenged the official propaganda. Contemporary online Chinese nationalism was first embarked upon in 1999 after the bombing of the Chinese embassy in Belgrade; the precedent of the Strong Nation Forum was exactly the "BBS Bulletin Forum Protesting Against the Violence of NATO." The second wave can be regarded as building momentum in 2005, when the first online nationalistic campaign was mobilized into physical movements against Japan. As various communicative theorists argue, speedy communication has been crucial for the creation of this new form of Web 2.0 nationalism, showing that the netizens can be anti-Western on one hand while still being critical of the party-state.[25] As Shih-Diing Liu has commented, such nationalism is different from the officially propagandized nationalism; it is bottom-up in nature and has emerged from the people (*renmin minzu zhuyi*).[26]

The Scope of the Book

Inevitably, a book such as this will have methodological limitations. Even though each author is an expert in his or her relevant geographical or policy area, it is clearly not possible to cover all aspects of contemporary Chinese online nationalism in one publication. The contributors can only dedicate themselves to identifying one or two representative perceptions constructed by Chinese netizens about a particular area or region. The contributors to this book, however, have always sought to address the following questions:

1. Which are the most important Internet sites carrying online discussion of Chinese nationalism related to the author's particular area of study?
2. What are the differences between online Chinese nationalism and the conventional form of nationalism? Why are there these differences?
3. Has nationalist online expression influenced actual foreign policy-making?
4. Has nationalist online expression influenced discourse in the mainstream mass media in China?
5. Have there been any counterreactions toward online Chinese nationalism? Where do they come from?

The book is divided into three sections. In the first, Sow Keat Tok provides the theoretical background to the analysis of online nationalism, focusing extensively in chapter 2 on the interrelationship between online Chinese nationalism and China's sovereignty. The second section explores the "first layer" of online Chinese nationalism, that is, the responses toward the United States, Japan, and Taiwan. In chapter 3, James Reilly analyzes the online Chinese response toward Japan from 2002 to 2005, concluding that the widespread use of the Internet in China has worked against the liberal assumption on cyberpolitics and in fact has strengthened emotional anti-Japanese sentiments. The analytical line is further picked up by Shih-Diing Liu in chapter 4, which places special focus on the 2005 demonstrations and the aftermath of their spillover effect. In chapter 5, Simon Shen provides a comparative view of Japan with his study of the diasporic contribution made to online Chinese nationalism. He uses Hong Kong as the case study, arguing that online comments made about Japan in China's special administrative region are relatively apolitical and apathetic, and have been used more to distance the mainland identity than to embrace it. In his innovative method and critical analysis, Benson Wong discovers in chapter 6 that online Chinese nationalism toward Taiwan has an ethnocentric perspective and has adopted a negative view of the political entity on the other side of the Strait. In chapter 7, Yaling Pan presents an analysis of online nationalism toward the United States by featuring a nice overall mixture of the mainland Chinese perspectives.

The third section of the book includes what we identify as the "second layer" of online Chinese nationalism, that is, the responses toward other geographical regions in the world. By referring to it as the second layer by no means undermines the importance of these regions: the fact that they have been little studied adds significant value to their being featured here. Starting with China's immediate neighbors, Chun Zhang in chapter 8 focuses on the online Chinese view toward Southeast Asia and discusses how ideational attachment and pragmatic concerns are combined in the cyberworld. In chapter 9, Winnie King describes the online Chinese nationalist response to the European Union over the disruption of the Olympic torch relay in continental Europe in 2008. Tied to the study of the European Union is the online Chinese nationalist response to the United Kingdom, perceived as the first nation in modern history to have humiliated China. In chapter 10, Chun-wing Lee discovers that in the eyes of Chinese netizens the United Kingdom has a dual identity, as both a special partner of the hegemonic Americans and a European power. To conclude this section, in chapters 11 and 12 Simon Shen and Kai-chi Leung investigate the online Chinese perception of Africa and Latin America, providing one of the forerunning academic studies on the interaction between online Chinese nationalism and these regions of the developing world. The findings of each chapter are summed up and analyzed in chapter 13, which discusses the implications of online Chinese nationalism toward Chinese foreign policy in general.

Notes

1. Message ID 4905226, *Paopao Julebu* (Bubble Club), 16 January 2009 [pop.pcpop.com/zpt/default.html?MainUrl=pop.pcpop.com/090116/4905226.html, accessed on 20 June 2009].
2. "Gongye he Zixunhuabu Xinwen Fayanren jiu Luse Shangwang Guolu Ruanti Wenti da Jizhewen (The Response from the MIIT Spokesman to the Questions Raised by the Press on the Issues of Green Online Filtering Software)," *Xinhua News*, 30 June 2009.
3. Office of the United States Trade Representative, "Secretary Gary Locke and USTR Ron Kirk Call on China to Revoke Mandatory Internet Filtering Software," 24 June 2009.
4. "Zhuanzai: 2009 Niming Wangmin Xuanyan (Forward: the Pledge of Anonymous Netizens)," *Hong Kong Inmedia*, 26 June 2009.
5. China Internet Network Information Center (CNNIC), *Statistical Survey Report on the Internet Development in China*, January 2009.

6. Andrew Chadwick, *Internet Politics: States, Citizens and New Communication Technologies* (New York: Oxford University Press, 2006).

7. Andrew Chadwick and Philip N. Howard, "Introduction: New Directions in Internet Politics Research," in Andrew Chadwick and Philip N. Howard, eds., *Routledge Handbook of Internet Politics* (Oxon: Routledge, 2009), 1–10.

8. Bingchun Meng, "Regulating *e gao*: Futile Efforts of Recommendation?," in Xiaoling Zhang and Yongnian Zheng, eds., *China's Information and Communications Technology Revolution* (Oxon: Routledge, 2009), 52–67.

9. Guobin Yang, "The Internet and the Civil Society in China: a Preliminary Assessment," *Journal of Contemporary China*, Vol. 12:36 (2003), 453–75.

10. Jürgen Habermas, *The Structural Transformation of the Public Sphere: An Inquiry into a Category of Bourgeois Society*, Thomas Burger, trans. (Cambridge, Mass.: MIT Press, 1989), 27–56.

11. Yongnian Zheng and Xiaoling Zhang, "Introduction," in Xiaolong Zhang and Yongnian Zheng, eds., *China's Information and Communications Technology Revolution*, 1–16.

12. Guobin Yang, "Historical Imagination in the Study of Chinese Digital Civil Society," in Xiaolong Zhang and Yongnian Zheng, eds., *China's Information and Communications Technology Revolution*, 20.

13. Yang, "Historical Imagination," 22.

14. CNNIC, *Statistical Survey Report on the Internet Development in China*, January 2009, 27–29.

15. Sun Zhigang died in 2003 after being beaten by police. It is thought that he argued with the police that detained him for not carrying his *hukou* ID card. In Harbin in 2003, a woman driver of a BMW apparently deliberately ran over and killed the wife of a tractor driver who had scratched her car (possibly having previously attacked the tractor driver and his wife), but only received a minor and suspended sentence for what was deemed a traffic accident. Liu Yong was the leader of a criminal gang in Shenyang whose original death sentence was reprieved as his confession might have been obtained through torture—though the reprieve was thought by some to be a result of his close connections with officialdom. He became the first person to be retried in China for an ordinary criminal case, and was immediately executed after the second trial passed the death sentence. For details, see Guobin Yang, "Historical Imagination in the Study of Chinese Digital Civil Society," 22.

16. Yongnian Zheng, "The Political Cost of Information Control: The Nation-state and Governance," in Xiaolong Zhang and Yongnian Zheng, eds., *China's Information and Communications Technology Revolution*, 141.

17. Thomas Christensen, "Chinese Realpolitik: Reading Beijing's World View," *Foreign Affairs*, Vol. 75 (1996), 37–52.

18. David Campbell, *Writing Security: United States Foreign Policy and the Politics of Identity* (Minneapolis: University of Minnesota Press, 1992), 62.

19. Alexander Wendt, *Social Theory of International Politics* (Cambridge: Cambridge University Press, 1999), 225.

20. Gerard Delanty and Chris Rumford, *Rethinking Europe: Social Theory and the Implications of Europeanization* (London: Routledge, 2005), 51.

21. For example, see Gideon Rose, "Neoclassical Realism and Theories of Foreign Policy," *World Politics*, Vol. 51:1 (October 1998), 144–72.

22. Baogang He, *The Democratic Implications of Civil Society in China* (Basingstoke, Hampshire: Macmillan, 1997); Jonathan Unger, "Bridges: Private Business, the Chinese Government and the Rise of New Associations," *The China Quarterly*, No. 147 (September 1996), 795–819.

23. Christopher R. Hughes, "Interpreting Nationalist Texts: A Post-structuralist Approach," *Journal of Contemporary China*, Vol. 14:43 (May 2005), 247–267.

24. Peter Hays Gries, "Chinese Nationalism: Challenging the State?" 256.

25. Karl Deustch, *Nationalism and Social Communication: An Inquiry of the Foundation of Nationality* (Cambridge, Mass.: MIT Press, 1953); Ernest Gellner, *Nations and Nationalism* (Ithaca, N.Y.: Cornell Press, 1983); Benedict Anderson, *Imagined Communities: Reflections of the Origin and Spread of Nationalism* (New York: Verso, 1983).

26. Shih-Diing Liu, "China's Popular Nationalism on the Internet. Report on the 2005 Anti-Japan Network Struggles," *Inter-Asia Cultural Studies*, Vol. VII:I (2006), 144–55. See also Liu's later chapter in this book.

2

Nationalism-On-Demand?

When Chinese Sovereignty Goes Online

Sow Keat Tok

THE ISSUE OF CHINESE SOVEREIGNTY regarding its territorial claims regained prominence in the first half of 2008, following the outbreak of civil unrest in Tibet in March. The unfortunate overlapping of the Tibetan disturbances with other events, in particular with Beijing's poor public relations management of its Summer Olympic torch relay, as well as the convening of the annual meetings of the National People's Congress (NPC) and Chinese People's Political Consultative Conference (CPPCC) has incited strong Chinese responses across the board, from the political leadership at the very top, down to the lower strata of Chinese society.

This agitation was especially evident within the Chinese online community. Livid discussions took the Internet by storm, as Chinese language websites, Internet bulletin boards and blogs were swamped with emotional outpourings in defense of the honor of the Chinese nation. The Chinese online community adopted a siege mentality against the outside world, as Chinese "netizens"—estimated to number some 253 million by an official count[1]—locked arms against all accusations, true or false, lobbed at the Chinese state. The ensuing tones were, unsurprisingly and uniformly, anti-Western. Seemingly concerted efforts to attack Western establishments like French retailing giant Carrefour and media consortiums CNN and *The Times* were carried out in the name of the Chinese nation.[2] In the eyes of these Chinese netizens, the autonomy of the Chinese people as a nation was fundamentally challenged by the conceited West; China's sovereignty was at its deepest peril.

Or was it?

Whichever is the answer to this question is not the point of focus here; more intriguing questions are: how can we understand this close association between the new phenomenon called "online Chinese nationalism"[3] and the Chinese image of its sovereignty? How does it come about and evolve to its form today? As a political identity, what does being an "online Chinese" suggest? To whom does this online "nation" apply, and how can it be, or have been, manifested? And most of all, how has this form of nationalism interacted with the Chinese regime today?

Manuel Castells wrote, "It is easy to agree on the fact that, from the sociological perspective, all identities are constructed. The real issue is how, from what, by whom, and for what."[4] This chapter responds to this challenge, so as to set the stage for further in-depth discussions in this volume on cases of online Chinese nationalism. Discussions on the relationships between online Chinese nationalism and Chinese sovereignty are carried out in three separate sections in this chapter. On the premise that online Chinese nationalism is an evolutionary product of its non-online counterpart brought about by technological advancement, rather than technology alone, the first task will be to outline the relationship between Chinese sovereignty and modern Chinese nationalism. The first section is thus dedicated to tracing the roots of the latter in China's historical and political experiences with sovereignty. An imported concept from the West, "Chinese sovereignty," or the awareness of it in the Chinese minds, was but a contingent product of the struggles that characterize China's political scenes from the late Qing dynasty until today.

The second section places this historical legacy in the current context. The Internet age came about in China at the same time as political conservatism regained prominence in Chinese thinking in the post-Tiananmen years. Unsurprisingly but not the least interestingly, the profile of an average member of the Chinese online community today coincides with that of socialization and upbringing befitting such a political and intellectual climate. This section discusses those important events that have led to the evolution of online Chinese nationalism in the past decade.

The third section then assesses the implication of online Chinese nationalism—or "patriotism" (*aiguo zhuyi* or *aiguo sixiang*) as officially known in China[5]—toward Beijing's policy-making process, especially with respect to Chinese sovereignty. Despite its relative nascent form, online Chinese nationalism has undergone at least three developmental stages. In each stage, this online sentiment has manifested itself primarily in different ways, and against different "others." These accounts provide evidences to the erratic, diverse, and sometimes conflicting, natures of such emotional outpourings over the Internet, against which Beijing has to constantly adjust its foreign as well as domestic policies. But can it possibly spin out of Beijing's control? Given online Chinese nationalism's ambivalence toward the official patriotic discourse in mainland China, its limited agenda and whatever little agreements achieved between its advocates so far, this author raises serious doubts about such prospects. Indeed, sometimes it could be more helpful to regard online Chinese nationalism as an offshoot of the official stance on patriotism: while having a life of its own, online Chinese nationalism appears to be operating within the framework in which it was born. More importantly, Beijing has so far shown enough tact to keep online nationalism on tab—leaving it simmering but not boiling over.

To round things up, through these discussions, this chapter seeks to problematize the idea of "online Chinese nationalism" by raising the following points: first, while the Chinese authorities have increasingly relaxed their control on the Internet in China today, there are lingering doubts about the extent to which this relaxation has made substantial impacts on issues related to Chinese sovereignty. This is in large part due to the structural constraints imposed by the Chinese historical memories surrounding China's experience with sovereignty as much as the state's continuing interests and efforts at maintaining such memories. Second, the result of such structure is the creation of a relatively unified community of online "nationalists" whenever the sovereignty of the Chinese state was deemed threatened. The structural constraints of Chinese nationalism denote that such nationalistic expressions, even when intense, hardly place issues outside the thinking box. This leads to the view that it might be necessary to ontologically separate the domestic from the international to understand this phenomena called "online Chinese nationalism." It is a complex social movement which cannot be dissociated from its non-online counterpart. It has historical roots that a self-professed "Chinese nationalist" would identify; moreover, it is grounded in contexts that reinforce the image of self, its contemporaries, and the world outside; it is shaped by the political processes, past and present, to attain its recognizable form today; it is also a highly divided process that besides a small, central core on which most in the movement would concur, very little agreement exists elsewhere.

Due to its explanatory nature, the chapter has engaged a more descriptive, rather than analytical, approach. The emphasis on placing historical descriptions within the contemporary context is particularly noteworthy. Historians John K. Fairbank and colleagues have noted,

> East Asia can best be understood through its history . . . the peoples of East Asia, more than those of the rest of the world, see themselves in historical perspective. They are strongly aware of their heritage. To approach them through their history is to look at them as they see themselves. . . . Only as one looks at the

> long flow of East Asian History can one perceive the direction of its motion and understand what is happening there now.[6]

Located at the center of East Asia, "China" is an embodiment of its long historical legacy, and history will be where this inquiry begins.

Locating the Link between Chinese Sovereignty and Nationalism

Both sovereignty and nationalism are relatively modern concepts in politics. Sovereignty, or *souveraineté* (as well as comparable, older terms such as *majestas* and *summa potestas*), first appeared as a coherent concept in the writings of Bodin after the Augsburg settlements in the sixteenth century. Core to this is the "new" idea of *cujus regio ejus religio*, a principle which affirmed the new religious order whereby the choice of religion rested entirely in the hands of the ruler of the land. At that time when religion was pervasive in the social and political fabrics of Europe, this was translated to a reorientation of decision-making powers. By Bodin's definition, the true attribute of the sovereign is "the power to make law,"[7] and "the distinguishing mark of the sovereign is that he cannot in any way be subject to the commands of another. . . . A sovereign prince is not subject to the laws of the Greek, or any other *alien* power . . . except in so far as they embody the law of nature."[8] This definition, in more contemporary terms, refers to the "idea that there is a final and absolute authority in the political community . . . *and no final and absolute authority exists elsewhere.*"[9]

Likewise, "nationalism" is also a relatively new concept. While some scholars predate the idea to the Middle Ages and perhaps even earlier,[10] it is generally accepted that, as a modern political concept, "nationalism" was *popularized* only in the eighteenth and nineteenth centuries.[11] By "nationalism," this chapter refers specifically to the collective belief that the community's territory, its people and their historical memories, public culture, as well as their rights and obligations, are congruent to its political claims.[12]

Sovereignty and nationalism are also closely related concepts. Benedict Anderson, in *Imagined Communities*, argued that nations are imagined as political communities as much as they are imagined as limited and sovereign, the political entity we all known today as the *modern state* being the interlocutor between the two.[13] The development of the modern state over the past centuries cannot be dissociated from the development of both concepts: first, with the infusion and refinement of the concept of "sovereignty" into European church-state relations in the sixteenth and seventeenth centuries through the contributions of Machiavelli, Luther, Botero, Bodin, Hobbes,

Locke, and Grotius, among others; then the slow convergence of the *nation* and the modern *sovereign* state in the succeeding centuries, beginning with the establishment of the French and American Republics, and culminating in Woodrow Wilson's Fourteen Points, in which the idea of *national sovereign state* was epitomized.[14]

Importing Sovereignty and Nationalism into China

However, when imported into China as a "Western" concept of international politics, sovereignty manifests far greater meanings in its relations with nationalism than was originally conceived. Sovereignty (*zhuquan*) appeared in the Chinese lexicon in the mid-nineteenth century, yet over time it has become one of the integral ingredients in the construction of the nationalistic ideals in China. Until its appearance in the Chinese translated version of Henry Wheaton's *Elements of International Law* (*wanguo gongfa*) in 1864, the idea of "sovereignty" had no cognitive meaning in the minds of the Chinese.[15] The late Chinese suzerain order, put in place to regulate the relations between China and its surrounding regions for the previous two millennia, and then maintained by the Qing (Manchu) imperial court, had paid scant attention to the territoriality of the state.[16] Friedrich Kratochwil considered the Treaty of Peking between Russia and China in 1860 as the watershed mark which led to "a more precise definition of the relationship between the suzerains and Russia" (and the West in general), though he also noted that this new arrangement with the outside world effectively took decades before becoming more naturalized in the Chinese state of affairs.[17]

Hence, to put things in perspective, the date-stamp commonly attached as the birth of Chinese nationalism by common nationalist texts to the First Opium War (1839–1842) appears tenuous at best. The self-strengthening movement (*yangwu yundong*, also known as the *ziqiang yundong* or *tongguang xinzheng*) that characterized Chinese politics in the mid-nineteenth century was by and large a direct response to the threats to the integrity of the Qing imperial court and the supremacy of the Chinese world order, rather than to the Chinese "nation" as a whole. The deep ethnic fissures that ran between the Manchu rulers and their Han (and other) subjects deterred the formation of a coherent "nation" based within the Chinese imperial state.[18] All humiliations suffered as a result of the Western encroachments were presumably the failures of the Qing court, not of the Chinese nation.

It was after the First Sino-Japanese War (1894–1895) that "Chinese nationalism" took nascent forms.[19] During the war, the Qing government suffered its most humiliating defeats in history despite its earlier self-strengthening policies: the war was remembered (perhaps erroneously) as China's first defeat by an

Asian state (Japan)—rather than a Western power—located at the peripheries of the traditional Chinese world order, and which until the 1860s was a conservative, inward-looking tributary state to the Qing court. Possibly nothing else could have come as a greater shock to the Chinese mentality then. In Jonathan Spence's words, the Chinese defeat "was a dark conclusion to the brightest hopes of the era of self-strengthening," as the Qing court's "sovereignty" over the island of Taiwan, the Pescadores, the Liaoning region, along with the Chinese influence over the Korean peninsula, were ceded to Japan following the war.[20] This defeat accelerated the demise of the Qing court and paved the way for the birth of the Chinese Republic.

Structuring the Relationship between Chinese Nationalism and Sovereignty

The divided Qing court responded to the crisis in a haphazard manner befitting an empire in decline. The young emperor Guangxu embarked on the short-lived Hundred Days' Reform (*wuxu bianfa* or *bairi weixin*) in 1898, only to be thwarted by the conservatives, led by the Dowager Empress Cixi. The conservatives attempted to reverse the waning fortunes of the Qing court by placing their trust in the Boxers Movement, which in the end turned out to be yet another disaster. At the same time as the Qing court and Boxers were creating mayhem, another storm was brewing among the country's intellectual elite. Following China's defeat by Japan, theories of social Darwinism took hold of China's disaffected intellectuals.[21] Fear for the perceived "unfitness" of China inspired some to initiate entirely new intellectual and political enterprises. "Nationalism" (*minzu zhuyi*) first seeped into the Chinese discourse at the turn of the twentieth century, again as a foreign import.[22] But as pointed out by Frank Dikötter, this nationalism exhibited more characteristics of "race" than "nation," sovereignty being a far less of a variable in the equation than anti-foreign sentiments.[23]

It was Sun Yat-sen who eventually found the necessary political expressions for Chinese nationalism in his writings and speeches. These ideas, more notably known as the "Three Principles of the People" (*sanmin zhuyi*), were developed and refined over a period of two decades, from 1904 to the mid-1920s. First and foremost of the three principles outlined by Sun was "nationalism." Given Sun's relative failure at capturing the imagination of the Chinese people in the earlier days of his revolutionary endeavors, it was no real surprise that "nationalism" became the foundation of his political manifesto. In his attempt to debase the Qing court, Sun took great pains to shift the discourse of political legitimatization in China from the traditional "mandate of heaven" to appeal to the modern, national cause. His previous calls to "drive out the Manchus to revive the Chinese rule" (*quchu dalu, huifu zhonghua*) were ridden with a strong sense of ethnic Han bias, but practical concerns drove Sun to take a more conciliatory approach. In the later years of the nationalist revolution he moved to include other ethnic minorities in the Chinese nation.[24] Whether or not Sun's rhetoric of Chinese nationalism *actually* spurred the eventual formation of the Chinese Republic in 1911 remains an open question, and perhaps best left to historians. But as a matter of interest to readers, one version of interpretation by a Western academic went so far as to conclude that "the true face of the 1911 Revolution belonged to Yuan Shikai, not Sun Yat-sen," as he argued that the success of the 1911 revolution probably owed as much to the disaffected Chinese elites and Qing military commanders as to Sun's visionary goals for the Chinese nation.[25]

The point of contention here is that Sun's Chinese "nation," at best, was a contingent product of the intense political competition in the early twentieth-century pre-Republican China. Central to his call to reorganize the Chinese peoples into a coherent nation, according to Sun, was the impetus "to save the country" in a world regulated by Darwin's principle of natural selection:

> We [the Chinese people] are the poorest and weakest state in the world, occupying the lowest position in international affairs; the rest of mankind is the carving knife and serving dish, while we are the fish and the meat. Our position now is extremely perilous; if we do not earnestly promote nationalism and weld together our four hundred millions into a strong nation, we face a tragedy—the loss of our country and the destruction of our race. To ward off this danger, we must espouse Nationalism and employ the national spirit to save the country.[26]

And "saving the country" involves political mobilization via the nationalist cause to preserve China from "dismemberment" and the loss of her sovereignty. Sun drew out the glories of the Chinese nation against the shame of being in an international position worse than that of its previous territorial holdings Annam (Vietnam) and Korea, to kindle the shame of the Chinese nation: "[w]e taunt the Koreans and Annamese with the name '*wang-kuo nu*' (slaves without a country), yet while looking at their position we seem unaware that our own position is lower than theirs . . . China is not the colony of one nation but of all, and we are not the slaves of one nation but of all. . . . We are not a semi-colony but a hypo-colony."[27] Territoriality and national sovereignty were pivotal in Sun's survey of potential threats against the Chinese nation, and to prove his claim, he (justifiably) turned to the long list of foreign exploits on a weak, Qing-led China in the preceding century.[28]

This shame-inspired nationalism, definable in the context of Chinese sovereignty, is the foundation of all political mobilizations in China for the succeeding century.[29] After the Nationalists were swept into power in 1911, if only to be cast aside by the conservatives and warlords mere months later, continuing division of the state and threats from foreign powers (especially Japan) set the tone for the build-up to the May Fourth Movement in 1919. It was during the latter event that Sun's rhetoric of national salvation captured the imagination of both the intellectuals and political elites alike. It was then that Chinese nationalism appeared to reach its zenith form, so to speak;[30] the structure of the relationship between the Chinese nation and its sovereignty, hitherto in flux, contested, and contingent, was finally cast in stone.

While Sun and his nationalistic ideals were not unique to himself—his contemporary intellectuals and activists like Liang Qichao, Yan Fu, Zou Rong, and so on argued along similar lines when mobilizing for "national salvation" and revolution—these ideas were eulogized in the succeeding regimes of Chiang Kai-shek and Mao Zedong as powerful precursors to shore up their respective political legitimacy.[31] John Fitzgerald may have overlooked one obvious fact when he made the claim that China is a "nationless state," as in "the Chinese nation has been created and re-created in the struggle for state power, and it has ultimately been defined by the state as a reward of victory."[32] The fact is that the various political constructs of the Chinese nation at different eras in the past century have exhibited far greater similarities in terms of their basic building blocks than Fitzgerald has perceived: foreign imperialisms and their encroachments of China's sovereignty *must* be countered at all cost; the glory of the nation *must* be restored, beginning by building up the required capacity to keep the nation and its holdings, especially territories, together and resist counterforces aimed at breaking up the country; "China" and its nation is constantly in danger of being wiped out from this realist world, and it is necessary for its people to remain vigilant at holding it together. These persistent efforts to leverage upon the humiliations and crises of the Chinese nation contributes to one of the most noteworthy phenomena of modern Chinese history—the strange, continued passion of the Chinese mentality toward the fascination of the "sovereign whole" and the messianic devotion to preserving it.[33] These themes of Chinese nationalism were constructed and reconstructed, or, to use Anthony Gidden's term, "structurated," constantly.[34] Chiang exercised this approach in his northern expedition, the war against Japan, and the civil war; Mao did likewise in the Chinese Communist Party's bid for power over the mainland, and also subsequently in his foreign policy and in the numerous domestic social engineering campaigns initiated from the 1950s to the 1970s.[35] Today, despite decades of separation and seismic socioeconomic changes, Chiang's and Mao's political successors across the mainland and Taiwan remain steadfast on their sacred mission to "unite" the state, the nation, and its territories under one single banner.[36]

Contextualizing *Online* Chinese Nationalism

The chapter has so far attempted to locate Chinese nationalism vis-à-vis its sovereignty from a historical perspective. What, then, is the relation between *online* Chinese nationalism and sovereignty? It is easy to conclude by simple extrapolation that the online version of Chinese nationalism is a mere function of its historical setup. However, to effectively capture the dynamic nature of such online sentiments, as well as to achieve a better understanding of the online nationalist movement in China today, it is only prudent to take into account the contexts in which Chinese nationalist sentiments are embedded today.

To begin, the latest report by China Internet Network Information Center (CNNIC) offers some interesting statistical insight to the structure of the Chinese Internet population, their common user profiles, as well as their patterns of usage, summarized below (all as of June 2008):

- In China, the overall penetration rate of the Internet was 19.1 percent. This represented less than one-fifth of China's official population count. In comparison, the global penetration rate of the Internet averaged 21.1 percent, and China's figure was ranked well below that of its neighbors like Korea and Japan, which stood at 71.2 percent and 68.4 percent respectively.
- Youths between the ages eighteen to twenty-four formed the largest group of Internet users in China, representing 30.3 percent of the total; 19.6 percent were under the age of eighteen, while those aged twenty-five to thirty took up 18.7 percent. Put together, these three age groups accounted for more than two-thirds of all Chinese Internet users.
- Students formed the largest group of Internet users by profession, at 30.0 percent. This was followed by employees (in nonmanagement positions) in public and private enterprises, at 25.5 percent, and self-employed and "freelance" workers, at 19.7 percent.
- Internet access from home had become a more common phenomenon, and at 74.1 percent was way ahead of accesses from other locations. Correspondingly, the total number of personal computers that were connected to the Internet reached 84.7 million.
- Access to, and publishing on, web media and Internet communities was becoming more common as well. For the first time, visiting blogs/personal spaces and BBS/Internet communities were ranked amongst the ten most common Internet activities. There were more than one hundred million blog/personal space owners, 28.0 percent of whom were updated at least once every six months.[37]

To put things in a more descriptive manner, the average Internet user in China today is most likely to be below the age of thirty (i.e., born after 1980), who is either still schooling or at an early stage of his/her career; he/she usually will access the Internet from home, frequenting new media websites, personal blogs, and the Internet communities through accessing, editing, and publishing activities. For all that having been said, these users represent only a fraction of China's total population: less than one out of five have access to the Internet; even fewer are actively engaged in Internet publishing and online opinion-building activities.

The above descriptions offer invaluable clues to the context in which online Chinese nationalism has developed over the past decade and its relationship with sovereignty. A closer examination of the formative years, through the 1980s to the early 2000s, of these age groups could direct us to the crux of the issue.

Legitimatization Crises of the 1980s and Early 1990s

More than a handful of notable scholars on Chinese politics have observed the trend of rising nationalism following Deng Xiaoping's Open Door policy in the late 1970s. Among them, many have argued that the Chinese Communist Party (CCP) has, at various stages since 1949, attempted to boost its flagging political legitimacy by stressing its nationalist credentials to offset the erosion of its ideological appeal.[38] This explanation certainly has its merits. Were Communism to be regarded a religious belief, the CCP's affair with capitalism possibly constitutes one of the deadliest, if not *the* deadliest, sins it has ever committed. Communism is no religious belief, of course, but the ideological communion of the two polemics is no less lethal to a communist party's stranglehold onto power. In this light, the success of Deng's turn toward capitalism was a miracle, not just in terms of its results, but of the double paradox that the party has survived the ideological crisis it has initially created for its political perpetuation. Encapsulated within Deng's aphorisms for pragmaticism and glory in China's search for economic development, for example "to get rich is glorious" (*zhifu guangrong*),[39] the frenzies and fervors of the Cultural Revolution were toned down and traded for more controlled and objective-minded pursuit of economic well-being.

The events of 1989 that led to the Tiananmen incident, the end of the Cold War, and the subsequent reversion of the CCP to the conservative ways brought about a shift in policy focus in the early 1990s. The ways of the pragmatists were moved aside for more dogged emphases on "patriotism," which in effect was a reintensification of nationalist education. Through his analyses of China's patriotic education campaign in the early 1990s, Suisheng

Zhao described this as the process in which nationalism was "rediscovered."[40] Indeed, up until the mid-1990s, *guoqing jiaoyu* (state of the nation education) and patriotic studies were introduced to all levels of education in China, while college-bound youths were subject to further indoctrination through a mandatory one-year national service in the guise of "military" training, but which in reality amounted to little more than physical training coupled with regimented and enforced political education.[41] A central document, "The Outline for Conducting Patriotic Education," was issued by the CCP Central Committee in 1994, at the apex of this wave of patriotic education. The purpose, again according to Zhao, "was to help the young people understand where China was strong, where it lagged behind, and what were its favorable and unfavorable conditions so as to enhance their sense of historical mission and responsibility."[42]

The reception for this state-led nationalism was apparently quite favorable at the societal level, which was already simmering with nationalist revivalism since the 1980s. The Tiananmen students were rallying under the flag of patriotism when they turned out in hordes before the CCP leadership decided to put them down forcefully; similarly, the Chinese intellectuals had been lending themselves to the nationalist, strong state discourse as an alternative to the anti-traditionalist thinking in the 1980s;[43] meanwhile, nonfiction writers popularized the series of books centred upon the theme "China can say no," as popular fictions plunged patriotism and/or nationalism into whole new perspectives.[44] Hidden in the authors' discontents against the regime (for the latter's lack of national assertiveness) nonetheless are hopes that the party would lead the way in announcing the return of the Chinese nation on the international stage.[45]

Verifying Learning Experiences: Heightening International Tensions in the 1990s

The Internet generation were brought up in this zealous atmosphere that combined the powerful top-down as well as the bottom-up pushes for nationalism. Furthermore, these youths have found that their learning experiences were enhanced by the unfamiliar strategic environment of the post–Cold War world. No doubt, these experiences were further distorted and manipulated by the state propaganda machineries, which, despite enormous changes that have taken place in the Chinese media sector, remained powerful influences even until today.[46]

"Nationalistic feeling reaches zenith in times of crisis," claimed Dingxin Zhao when he wrote about the nationalistic zeal in Chinese students during the Belgrade bombing events in 1999.[47] For the whole decade in the aftermath of Tiananmen and the dissolution of the Soviet Union, China and its people were persistently portrayed by the state media as a crises-ridden nation, in line with the state's patriotic education campaign. These efforts thrust the CCP to the fore of the Chinese patriotic front, making it the guardian angel of the nation's pride and honor, while spurring its target audience to follow its strides.

The CCP's tactics found their corresponding illustrations in the heightened international tensions facing China in the 1990s. China's status as an international exile after Tiananmen helped to drum up this image of a deprived nation, its glory and honor once again denied by external (Western) powers. In particular, China's plight during the Opium War was opportunistically revived to rekindle the nationalistic sentiment; on one occasion, a symposium marking the one hundred fiftieth anniversary of the war was held on the day before the first anniversary of Tiananmen, on 3 June.[48] Furthermore, the end of the Cold War appeared to have made this image all the more believable as the United States, being the only superpower left standing, began to adopt an antagonistic stance toward China following the collapse of the Soviet Union. The euphoria in the liberal West in the early post–Cold War years put the CCP as one of the very last (and the most visible) oppressive regimes in the world. All of a sudden, as if everything so good turned out so bad, China found itself at the center of an international siege.

Earlier, Deng advocated a policy for China to "lie low and bide its time" (*taoguang yanghui, yousuo zuowei*). Under the tutelage of the paramount leader, China had endured difficult "exile" years to focus on its belated economic reforms. But Deng's successors were presumably less keen for such subtlety. China's failed bid for the 2000 Olympic Games was met with massive disappointment domestically, after a protracted build-up of its people's expectations.[49] In the run-up of its bidding process, Beijing endured international taunts on its human rights record while constantly reminding the nation of the significance of a successful bid. China's catastrophic failure to secure the 2000 games—Beijing had led comfortably in all voting rounds, and lost out to eventual winner Sydney only in the final round—cannot but instill an aura of victimization and international conspiracy within the expectant nation.

But China's failed bid was a mere blip in the grand scheme of events in the 1990s, as China continued to meet obstacles in its attempt to break its diplomatic deadlock with the rest of the world. Human rights abuses and continuing linkages to weapons programs in rogue states became customary accusations from the outside world. The "China threat" and other theories pointing at an imminent strategic standoff between China and the West emerged. All these complications only served to strengthen the impression of yet another international conspiracy against the Chinese nation.[50]

Chinese nationalists' frustration imploded when the United States purportedly showed its willingness to support a Taiwan "separatist" agenda during the Taiwan Strait crisis in 1995–1996. Responding to Taiwan's budding democratizing movement, the United States maneuvered its carrier battle groups near the Straits after the Chinese government conducted missile tests and military exercises targeting the island. For fear of discrediting its claims to safeguard China's sovereign unity, Beijing was forced to play hardball. It had to keep raising the stakes against Taiwan's perceived indigenous movements and U.S. intervention into Chinese sovereignty, while waiting for the other to blink in a game of bluff. In the passionate atmosphere that it had whipped up, the CCP stood to lose more than just a breakaway island in the game; the nationalists' anger could easily turn against the party should it fail to show resolve on the matter.[51] The danger of a real conflict was eventually averted, and the CCP emerged from the event smelling sweet of its national assertiveness, but at the same time the stage was set for the nationalist agenda: the egos of the Chinese nationalists were fed to an unprecedented level.

The Rise of China

A rising China is a deeply entrenched image in the contemporary mind-set. "The only thing rising faster than China is the hype about China," Meixin Pei once quipped.[52] Skeptics apart, in truth China actually did achieve phenomenal economic growth after coming out of hiatus in 1992; its economy steamed ahead even in the wake of the Asian financial crisis in 1997. The country has since become one of the world's top FDI destinations and largest exporters. China's foreign reserves are today second to none in the world, a staggering US$1.5 trillion.[53] Chinese homegrown corporations and sovereign funds have meanwhile reached out to the world, transforming the country's excess current account balance to economic power. Whether or not China has made sufficient headway in other realms remains to be ascertained, but imagine the significance of these remarkable advances in the eyes of a Chinese nationalist brought up under the CCP's patriotic education program!

Expectations of the nation continue to be high, and quite deliberately made so. There were clear attempts to continually lift the hopes of the nation. The retrocession of once-ceded territories Hong Kong and Macao in the late 1990s were *excessively* packed with nationalist tones.[54] Beijing at last securing the hosting rights to the 2008 Olympics was portrayed as the rebirth of the long-lost nation,[55] and China's long-overdue membership in the World Trade Organization was celebrated as a national event,[56] while the successes of manned space missions have been used to generate the kind of euphoria that helps to cement national unity.[57]

These efforts were similarly reflected at the leadership level across the last two decades. Arguably, Deng's plea to "bide one's time" has its hint toward a glorified national self. Deng's successor Jiang Zemin advanced the "comprehensive power" and "gearing with the world" ideas when China was emerging from its post-Tiananmen reclusive years; the task on his hands was to put China back on the map of international politics. With China's rise all but taken for granted, China's fourth-generation leaders under Hu Jintao spoke of China's "peaceful rise" (*heping jueqi*, and later "peaceful development" [*heping fazhan*]) and "going out" (*zou chuqu*).[58] The messages of these leaders were crystal clear: under the CCP's leadership, the Chinese nation has slowly, but surely, climbed out from the lowest point of its five-thousand-year history; the nation has finally come of age at the dawn of the twenty-first century.

Liberalization, Diversification, and Individualization of the Media

During the time of growing Chinese national pride in the 1990s, a fourth element provided the means without which online sentiments would not have blossomed. Since the mid- to late 1990s, rapid commercialization and liberalization of the media has introduced an entirely new dimension to the formation and expression of public opinion. Market-oriented reforms in the Chinese press industry had begun as early as during the 1980s when press companies were granted financial independence with limited editorial freedom.[59] Meanwhile, controls over other types of media companies, like the radio and television broadcasting stations, were loosened gradually, though it took China's WTO accession in 2001 to achieve full commercial liberalization.[60]

The Chinese traditional media market is *huge* with its enormous viewership, but despite the fact that the pie is large, the competition is no less intense. By the early 2000s, not counting the mainstream newspapers, more than 800 independently established newspapers were circulating in China; registered journals and periodicals of all natures numbered some 8,725 across the board; while its broadcasting and televising industry was probably the largest in the world—there were a reported 2,216 broadcasting and television stations, spread across China's county to national levels.[61] Drives to gain greater share of the viewership, advertising and broadcasting markets, and the new threats imposed by the introduction of competition from foreign media companies, promoted the growth of indigenous, at times semi-official, media "mega-conglomerates" like China Media Group at the national level, and other provincial-turned-national entities like Golden Eagle Broadcasting System (or Hunan Media Group) and Shanghai Media and Entertainment Group. The party's propaganda organs are coming under increasing pressures to bend to market demands to provide more varied, timely, and accurate reports. Party mouthpieces like the *People's Daily* and the China Central Television (CCTV) were literally forced to undergo further reforms and restructurings amidst decreasing viewership and market diversification.[62] News coverage and editorial opinions have also diversified, as the media jostle over their prospective audiences. Naturally, the winners of this game of wits and bravado (under the ever watchful eyes of the still-powerful Department of Propaganda) will be the ones who decide on stories or reports which resonate best with their viewers.[63]

Another development in the media industry is the rise of the so-called new media in China. The onset of the Internet, cable and satellite television, and the mobile phone services have not only formed part of the media liberalization and diversification process, more importantly, they have enabled *individualization* and *customization* of the media unseen in China before. At the consumer level, an individual is no longer subjected to his/her geographic or social position, as one can avail oneself with a mix-and-match choice of his/her sources of information, as well the methods to access them, based entirely on one's capacity to consume. More than that, these new media have revolutionized the idea of "journalism" altogether: personal blogs, MySpace, Internet chatrooms, short-messaging services (SMSs), and multimedia messaging services (MMSs) have altered the structure of information flow in the society. Any individual with his/her electronic gadget can now practically publish *anything, anywhere*, and *anytime*. The contents of the messages are not subjected to any censure—at least not until the discovery of certain "offensive" materials by the webmasters in the case of Internet contents, and perhaps not at all in other cases like mobile services. In other words, practically everyone in China today can be his/her own journalist if one chooses to, and this endeavor is not subjected to any predetermined journalistic code of conduct.

Consequently this created the necessary space for the explosion of information and opinions in China. This has a direct bearing on the expressions for nationalism. With the atomization of nodes of information, the construction and expression of the national self became just as individualized and diversified. The national opinion-maker ceased to be the private purview of the Chinese political leaders, for an individual can now emerge anywhere, from nearly any walk of live, to usurp that role. The Chinese top leadership will have to contend with sharing this public space with an ever more fragmentary discourse of "Chinese nationalism" originating from below, reaching in using increasingly varied, and advanced, forms of information and communications technologies. The CCP's monopoly on these matters has long since ended.

Nonetheless, it is also important not to read too much into such developments. Since the new media-types rest heavily on the dispersal of corresponding technologies, an individual's accessibility and technological know-how become overriding determinants for effective participation. It is easy to overlook the fact that online Chinese nationalism is a far greater phenomenon in the country's developed coastal cities and within the fairly well-educated, younger strata of the society. These groups possess greater propensity, as well as the adequate material capacity, to engage in the process; others, like the underdeveloped rural areas and those lesser-educated groups, have largely been left behind. Furthermore, to put things in perspective, the impacts of the ever-changing technologies have remained minuscule. Despite the increasing sophistication of technologies, searching for and accessing to Internet contents are still arcane sciences. The decentralized and time-sensitive natures of online information necessitate that, in order to pick out a particular set of information from the literally millions or even billions of other possible choices, an individual has to be *at the right place, at the right time.* The contingent nature of online content cannot be more emphasized here. Moreover, this difficulty works against the Chinese web censors as much as it works against potential users accessing the information. In truth, online Chinese nationalism has a far more limited reach than is generally portrayed.[64]

Nationalism-On-Demand?

In broad strokes, online Chinese nationalism can be divided into at least three identifiable stages, each preceded by either a single or a series of watershed events that brought online nationalism to another unprecedented level, against a different agenda and/or "enemy," and that have forced Beijing to scramble its state machineries to respond.

The first stage was marked by the 1999 U.S. bombing of the Chinese embassy in Belgrade, and characterized by the intense—and one may say, mutual—national animosity between China and the United States. This was also the period that first saw the entry of Chinese nationalism into the virtual space. During this stage, online China nationalism were founded on a strong anti-U.S. sentiment, brought about by what were perceived as U.S. attempts to identify a new, post–Cold War enemy in China, and blatant infringements of the Chinese sovereignty through a "containment" strategy aiming to isolate the rising giant. It is not uncommon to see online articles laced with nationalistic zeal in popular websites condemning U.S. post–Cold War interventionist policy, and conspiracy theories on U.S. "hegemonic ambitions."[65] Eventually, the tragedy of 9/11 abruptly shifted U.S. attention away from China toward international terrorism, and relieved Beijing of its hardline stance toward Washington, even though anti-U.S. sentiment has found a group of staunch supporters in the Chinese online community since.[66]

The second stage placed Japan as the next target for online Chinese nationalism. If the first stage was to be understood as a response to U.S. efforts to impede China's rise, this second, from 2003 to late 2008, can be seen as part of Chinese nationalists' rising confidence in China's role in world politics. From its participation in the international efforts against the "war on terror" to its WTO accession and various other successes in its economics, policy, and technological fronts, Beijing had given its nationalists a lot to cheer about. In the eyes of online Chinese nationalists, the time had finally arrived to address the long-standing historical issue between the two long-time nemeses in East Asia. This was catalyzed by Japanese Prime Minister Junichi Koizumi's political posturing, which was driven by domestic concerns.[67] Following infuriated responses from the Internet community on the mainland, relations between Beijing and Tokyo plunged to new lows.[68] Japan replaced the United States as the online Chinese community's all-time favorite for "national bashing"; outstanding sovereignty issues like Diaoyutai (Senkaku) Islands were put back into the spotlight, while China's purported submarine incursions into Japan's territorial waters were met with self-indulging enthusiasm and unfounded pride at some extremes.[69]

As this chapter is written, online Chinese nationalism is transiting into a third stage. Sino-Japanese ties gradually warmed after Koizumi stood down in 2006, and with China's successes in the 2008 Beijing Olympics, online Chinese nationalism is moving into a new domain. At this point, it is difficult to pin down the exact agenda, but assenting online responses toward the high-profiled visit by Chen Yunlin (the chief of the Association for Relations across the Taiwan Straits—ARATS) to Taiwan, and toward Beijing's assertive stance over the Tibetan issue suggest that the ground may be shifting yet again.[70] These online reactions may be hinting at a retreat from great power rivalries toward a form of new optimism and loud confidence over domestic sovereignty issues.

All three stages remind China observers of the remarkably fickle nature of online Chinese nationalism, and of the persistent demand on Beijing to keep close tabs so as to stay abreast with, if not ahead of, any subtle changes. Although it is an external variable that the CCP can gladly do without, the limited nature of online Chinese nationalism makes it an appealing ally to the political regime.

For a start, online Chinese nationalism is not only limited in its reach, its eyes are cast upon a limited agenda. If one assumes that online Chinese nationalism staunchly defends *all* issues of Chinese sovereignty, he/she will be in for a big surprise. On Chinese territorial sovereignty, online Chinese nationalism is fixated on issues like Taiwan, Hong Kong, Macao, Tibet, Xinjiang, and the Diaoyutai Islands, but generally lacks interest in, for example, the sovereign status of (Outer) Mongolia, the Russo-Chinese border, the Sino-Indian border, or the Sino-Vietnamese border issues, for that matter. At the same time as the Constitution of the Republic of China (Taiwan) continues to exert its sovereignty over Mongolia,[71] and an enthusiastic debate about a possible merger between the two states in *Qiangguo luntan* (State Strengthening Forum, an online forum under the *People's Daily*) in 2004,[72] Bejing seemed unwilling to change its "state-to-state" relations with its northern neighbor and its once historic territory. Likewise, since the 1990s, China has attempted to quietly stabilize its long-disputed land borders with Russia, India, and Vietnam through a series of diplomatic negotiations, and at the same time has made peaceful overtures toward the Spratly dispute.[73] These efforts, though riddled with tensions, have been "miraculously" spared of unwanted online attention domestically.

Online Chinese nationalists have been particularly reactive to China's relations with the big powers of the post–Cold War order, while they have paid scant attention on those more day-to-day exercises of China's sovereignty. In his interesting differentiation between "abstract" and "concrete" sovereignties, international legal scholar Wenhua Shan argued that China has always been more adamant in preserving the former, which is a legal, formal concept in international relations; meanwhile Beijing has shown more flexibility in the latter, the operational (concrete) aspect of sovereignty—"[i]n a strict sense, 'concrete sovereignty' is not 'sovereignty' per se, but subordinate powers or rights, which always imply sovereignty of which they only sanction the execution."[74] If one were to apply Shan's typology of sovereignty to our case today, evidence points to online Chinese nationalism operating along similar lines, though under different logic. Shielded from the operational and decision-making processes of Chinese sovereignty, but well-connected to the network of international news coverage, online Chinese nationalists are mesmerized by their nation's brushes with the world's greatest powers as well as its potential adversaries; this is a view much narrower than a casual observer may assume.

Operating from the structure of their imagination and understanding about the "Chinese history," these online nationalists exhibited little tendency to interpret issues outside that thinking hat. Amid Sino-U.S. tensions over the 1999 Belgrade embassy bombing incident, the 2001 spy plane incident, and the Tibetan unrest in early 2008, for example, Chinese nationalism that has been expressed online was consistent with the image of an aggrieved, yet proud, nation. Were one to turn the clock dial back a century earlier, these sentiments may come across as not that dissimilar to those expressed, again for example, in China's boycott of American products in the 1880s and early 1900s, or the explosive anger thrown up by the Chinese intellectuals and students during the May Fourth movement.[75] Thus, to put things simply, the lid of Chinese nationalism is capped tightly even when it goes online; and consequently, its sense of sovereignty is perhaps a contingent function of its historical memory. And this memory, this author would like to stress again, is a fairly consistent being—embraced and continually reinstated not only by different political regimes in China across time and space for the past century, but also within those Chinese communities now residing outside China.

From this vein, this author holds a rather pessimistic view of online Chinese nationalism.[76] To the Chinese political leaders, online Chinese nationalism poses inconvenience to policy-making, but not one that they cannot accommodate at the moment. Online Chinese nationalism's common roots with the official version of Chinese history make its outlook relatively pro-regime. Like a handful (one-fifth of the total population on the outside) of outspoken sailors on board a large sailing ship, online Chinese nationalists can occasionally be nuisances when Chinese political leadership chart the course of the state. This was the case, for example, when Beijing struggled to keep an even keel on the Japanese new history textbook issue in April 2005. Online nationalist sentiments no doubt contributed to the mobilization of the anti-Japanese movement then.[77] Similar cases can be made from the strong online opposition to Japan's bid for membership in the UN Security Council in 2005, and the dispatch of Japanese Self-Defense Force transport planes to assist in relief work in China's Wenchuan earthquake in May 2008.[78] Both actions may have eventually forced Beijing to quietly backtrack from its original plans, even if only to reinforce the CCP's position as the protector of Chinese sovereignty. Let there be no doubt that this cordiality can change in a whisk, from the moment that the party ceases to be seen as the "patriotic" front of the nation, but Beijing's extra effort, for the moment, appears worthwhile. Despite their contentious nature, these online Chinese nationalists are no challengers to the regime. Indeed, for all their articulated views and opinions, they appear to be still operating within the framework of nationalistic discourse directed by the regime; they remain loyal supporters to the ship that the CCP carefully designed, built, and commandeered.[79]

Continuing Search for Chinese Nationalism and Sovereignty

Aiguo wuzui ("patriotism is no crime"). As the saying goes in China, one can never be wrong by being too "patriotic" (or, in fact, "nationalistic"). To get a grip on this aphorism, it is crucial to first understand the role of nationalism and its relation with Chinese sovereignty since the mid-nineteenth century. As foreign imports, "nationalism" and "sovereignty" have been highly politicized by opinion-makers of all political persuasions throughout the turbulent times that the "Chinese nation" has come into being. Interestingly, this led to two contradictory developments. On the one hand, the persistent return to the same set of historical memories created a fairly consistent image of a nation in distress, which leaders of various political forces from Sun onward to Hu Jintao today have utilized to shore up their legitimacy. As a product of this structuring of historical memories and intensive patriotic indoctrination, Chinese nationalism represents a united and missionary effort to overcome the "enduring injustices," particularly the historical "sovereign deficit," owed to this proud people.[80] On the other hand, its politicized nature has given Chinese nationalism a rather elusive form. Its contested nature implies that, beyond its small, central core that calls to defending the honor of the nation is a large blurry region in which in reality no consensus exists. A self-professed Chinese "patriot" or "nationalist" from all political camps between the extreme left to the far right will be unlikely to dispute the historical injustices done to China during its transition to a modern state, though they will be less inclined to agree as to *whom* (i.e., who are the "Chinese"?) these injustices apply, *how* justice can be exacted, and to *what* ends this sentiment seeks.

Online Chinese nationalism may be no more than a spin-off from this age-old quandary, although it has found a new outlet through the new information technologies made available to Chinese consumers today. While this chapter will not go so far as to dismiss this emotional expression outright, several reasons require observers to be extra cautious when looking at online Chinese nationalism. This emotion is largely confined to a fraction of the whole "nation," comprising mainly of those IT-savvy youths in China's more developed cities. Given that the formative years of most in this age group fall into the years between the late 1980s and the mid-1990s, at the height of the Chinese central government's "patriotic education campaign," and during which China's international space was greatly constricted in the immediate post-Tiananmen days, it is not difficult to place this "new" nationalistic expression in context. With the introduction of consumerism, new information technologies like the Internet and other forms of mobile communications become more common. Together with the growth of a more consumer-based media industry, as well as the introduction of foreign competition to the erstwhile state-dominated sector, the pressures are now on the indigenous media companies to capture their ever fickle audiences. At the same time as the society locked into an insatiable cycle for more accurate and timely news reporting, consumerism also introduced a greater demand for sensational and controversial editorials.[81] These factors all rolled up to provide a culture in which online Chinese nationalism thrives.

While this form of nationalism has posed new challenges to the monopoly of the CCP on the discourse of the nation, it is also important not to overstate its influence. For a start, the decentralized nature of online materials works against its censors as much as against its target readers; unlike traditional published media which has a set of relatively fixed and quantifiable catchment groups, it takes more than a simple probability for online contents to reach their audience, let alone making it to among the top hits on Google or *Baidu* (a popular Chinese search engine). As a result, the popularity (*renqi*) of the content author—as in the cases of popular public figures and those so-called star bloggers—individualistic and alternative viewpoints, sensationalized phrasings and contents, and so forth, become important ingredients for a website's success among Chinese web surfers. Add in the contributors' relative youthfulness, a government-sanctioned Internet environment[82] and, more important, online Chinese nationalism's tunnel vision on national issues, and it remains doubtful if it forms a coherent force in Chinese political affairs.

For these reasons, online Chinese nationalism has largely been playing a supportive role vis-à-vis Chinese sovereignty issues. Even if there were occasional flashes of uncontrolled online movements, the state appears to remain well in control of the situation. The CCP has been playing its cards tactically with these online nationalists, trying to manage their somewhat intermittent outbursts of emotions carefully, and sidestepping on certain issues while carrying on its quiet diplomacy on others. Whenever the situation warrants it, Beijing has been careful to step back and put up a nationalistic front to quell discontents. Meanwhile, the regime is shrewd enough to keep the emotions simmering in the background so it can mobilize their support whenever required. Like the video-on-demand (VOD) technology so common to websites like YouTube and Google Video, online Chinese nationalism appears to fall into the same category: always ready on-demand, though the web engineers could just pull the plugs off the servers when the situation calls for it.

Notes

1. According to the CNNIC survey report released in July 2008, in the first half of 2008 alone, China's Internet users swelled by forty-three million, for the first time outstripping that of the United States (U.S.) in its total net population; by estimate, the Internet users in the United States totalled 230 million as of June 2008. China Internet Network Information Center, *Statistical Survey Report on the Internet Development in China No. 22* (Beijing: CNNIC, July 2008), 14–15.

2. "How Jane Macartney of *The Times* Became China's Most Hated Person," *The Times*, 29 March 2008; "Ugly Side of Online Nationalism in China," *Straits Times*, 26 April 2008. An interesting music video was made to taunt the Western media on their "bias reporting" over Tibet unrests; see "Don't be too CNN," YouTube [www.youtube.com/watch?v=uX28hMlkB5Y, accessed 15 August 2008].

3. Unless otherwise stated, by "China" and "Chinese" in this chapter, the author refers specifically to the referents of what is known as mainland China today. Admittedly other forms of "Chinese nationalism" exist elsewhere—those that were expounded in Taiwan and that amongst overseas Chinese communities were but two examples—and across time. This chapter is only interested in the online forms of nationalism that are primarily found on the Chinese mainland. In addition, the use of the singular form ("nationalism" rather than "nationalisms") does not suggest that this author sees Chinese nationalism, online or otherwise, as a unified and singular movement. In fact, this author believes that many versions of this "nationalism" coexist with each other, as will be unfolded as this chapter develops, and also taken up by other contributors in this volume. This is thus more of a consideration for grammatical consistency, rather than analytical discretion. Fu-Lai Tony Lu and Diana Sze Man Kwan, "Social Construction of National Identity: Taiwanese versus Chinese Consciousness," *Social Identities* 14 (2008): 33–52; Wang Gungwu, *China and the Chinese Overseas* (Singapore: Times Academic, 1991); Lynn Pan, *Sons of the Yellow Emperor: A History of the Chinese Diaspora* (New York: Kodansha International, 1994), chapter 13.

4. Manuel Castells, *The Rise of the Network Society*, second ed. (Oxford: Blackwell, 2000), 7.

5. "Patriotism" is commonly understood as the politically preferred term to denote nationalism in mainland China. There is a general acceptance of this terminology, even within academic circles. Some academics prefer the term "patriotism" to separate state nationalism from its popular form, while others have used this to set apart Chinese nationalisms geographically, that is, nationalisms in mainland China and elsewhere. This author, however, sees no need for such distinction to make his case in this chapter, and "nationalism" shall be used consistently throughout this discussion. See Yu-sun Chou, "Nationalism and Patriotism in China," *Issues and Studies* 32 (1996): 67–86; Phil Deans, "State Patriotism Versus Popular Nationalism in the People's Republic of China," *IIAS Newsletter* 34 (2004): 10.

6. John K. Fairbank, Edwin O. Reischauer, and Albert M. Craig, *East Asia: Tradition and Transformation* (London: George Allen & Unwin, 1973), 2–3.

7. M. J. Tooley, abridged and trans., *Six Books of the Commonwealth* (Oxford: Basil Blackwell, 1967), 43.

8. Tooley, *Six Books*, 28, 35 (emphasis added). See also Ingrid D. Delupis, *International Law and the Independent State* (New York: Crane Russak, 1974), 3.

9. F. H. Hinsley, *Sovereignty* (London: C. A. Watts and Co., 1966), 26 (emphases in original); also Hedley Bull, *The Anarchical Society: A Study of Order in World Politics*, third ed. (Basingstoke: Palgrave, 2002), 9. Bodin's definition of "sovereignty" was refined by later scholars over the succeeding centuries, though centrality of an ultimate authority over a limited "boundary" remains relatively unchanged. For Hans Morgen-

thau, as well as others who subscribed to the territoriality of sovereignty argument, this boundary is defined territorially or geographically; Anthony Giddens, Charles Tilly, and others would draw attention to the importance of coercion or violence elements in the authority over this boundary; yet others focused on the aspects external to this boundary, as that sovereignty is defined vis-à-vis "the others." In whichever case, this chapter shall not engage these definitions in depth but adopt Hinsley's authoritative definition instead. Hans J. Morgenthau, *Politics Among Nations: The Struggle for Power and Peace*, sixth ed. (New York: McGraw-Hill, 1985), 328; Daniel Philpott, *Revolutions in Sovereignty: How Ideas Shaped Modern International Relations* (Princeton, N.J.: Princeton University Press, 2001), 16; Anthony Giddens, *The Nation-State and Violence: Volume Two of a Contemporary Critique of Historical Materialism* (Cambridge: Polity Press, 1985), chapter 1; Charles Tilly, *Coercion, Capital, and European States, AD 990–1992* (Cambridge, Mass.: Blackwell, 1992), 19–28 and chapter 3; Alan James, *Sovereign Statehood: the Basis of International Society* (London: Allen and Unwin, 1986), 24.

10. See for examples, Hugh Seton-Watson, *Nations and States* (London: Methuen, 1977), chapters 1 and 2; Liah Greenfield, *Nationalism: Five Roads to Modernity* (Cambridge, Mass.: Harvard University Press, 1992); Adrian Hastings, *The Construction of Nationhood: Ethnicity, Religion and Nationalism* (Cambridge: Cambridge University Press, 1997).

11. See James Anderson's graphic presentation of the evolution of "nationalism" from late medieval times to the Second World War, James Anderson, "Nationalism and Geography," in *The Rise of the Modern State*, ed. James Anderson (Sussex: Wheatsheaf Books, 1986), 123. Also see Benedict Anderson, *Imagined Communities: Reflections on the Origin and the Spread of Nationalism* (London: Verso, 2006), 12, 192–93; also Elie Kedourie, *Nationalism* (London: Hutchinson, 1960); Eric Hobsbawn, *Nations and Nationalism since 1780* (Cambridge: Cambridge University Press, 1990).

12. This definition is a refinement that combines Ernest Gallner's statement that "[n]ationalism is primarily a political principle, which holds that the political and the national unit should be congruent" and Anthony Smith's "Western model of the nation," that it consists of a "[h]istorical territory, legal-political community, legal-political equality of members, and common civic culture and ideology." Ernest Gallner, *Nations and Nationalism* (Oxford: Basil Blackwell, 1983), 1; Anthony Smith, *National Identity* (London: Penguin, 1991), 11.

13. Anderson, *Imagined Communities*, 7.

14. In his Fourteen Points addressed to the U.S. Congress, Wilson used "nation" in place of the state, and in most of his speech, in particular point 5, he was in effect advocating congruence between the nation and the state—"in determining . . . questions of sovereignty the interests of the populations concerned must have equal weight with the equitable claims of the government whose title is to be determined." Woodrow Wilson, "Fourteen Points," in *Classic Readings of International Relations*, ed. Phil Williams, Donald Goldstein, and Jay Shafritz (Beijing: Peking University Press, 2003), 24.

15. Federico Masini, "The Formation of Modern Chinese Lexicon and its Evolution toward a National Language: The Period from 1840 to 1898," *Journal of Chinese Linguistics Monograph Series No. 6* (1993): 46–48.

16. Martin Wight, *Systems of States* (Leicester: Leicester University Press, 1977).

17. Friedwich Kratochwil, "Of Systems, Boundaries, and Territoriality: An Inquiry into the Formation of the State System," *World Politics* 39 (1986): 31.

18. John K. Fairbank, *The Great Chinese Revolution: 1800–1985* (New York: Harper Perennial, 1987), 18–21; Frank Dikötter, *The Discourse of Race in Modern China* (London: Hurst & Co., 1994), 20–30.

19. This point is shared by numerous scholars in the field. See, for example, Michael Yahuda, "Dimensions of Chinese Nationalism," in *Asian Nationalism*, ed. Michael Leifer (London: Routledge, 2000), 26–27.

20. Jonathan Spence, *The Search for Modern China* (New York: W. W. Norton & Co, 1991), 223–24.

21. Dikötter, *The Discourse of Race in Modern China*, 67–68, 101–2.

22. There are discrepancies to the exact year that the term "nationalism" appeared in writing. Dikötter put it at 1903, while Christopher Hughes cited another source to place it at 1901. Yet others argued that the term *Zhonghua minzu* (Chinese nation) appeared in writings as early as 1895. However, in any case the fact is inconsequential to the author's treatment of the development of the Chinese nation. See Dikötter, *The Discourse of Race in Modern China*, 97; Christopher Hughes, *Taiwan and Chinese Nationalism: National Identity and Status in International Society* (London: Routledge, 2000), 3; Han Jinchun and Li Yifu, "The Emergence of '*minzu*' in the Han Language and its Initial Usage" (*hanwen "minzu" yi ci de chuxian yiji chuqi shiyong qingkuang*), *Ethno-National Studies* (*Minzu Yanjiu*), 1984(2): 38–45.

23. Dikötter, *The Discourse of Race in Modern China*, chapter 4.

24. Frank W. Price, trans., *San Min Chu I: The Three Principles of the People* (Shanghai: The Commercial Press, 1928), 11–12.

25. Peter Zarrow, *China in War and Revolution, 1895–1949* (London: Routledge, 2005), 35. See also Michael Gasster, "The Republican Revolutionary Movement," in *The Cambridge History of China, Vol. 11*, ed. John K. Fairbank and Kwang-Ching Liu (Cambridge: Cambridge University Press, 1980), 527–29.

26. Price, *San Min Chu I*, 12.

27. Price, *San Min Chu I*, 38.

28. Price, *San Min Chu I*, 32–39, 122–48.

29. William A. Callahan, "National Insecurities: Humiliation, Salvation, and Chinese Nationalism," *Alternatives* 29 (2004): 199–218; Chou, "Nationalism and Patriotism in China," 67–77. See also efforts to conceptualize "humiliation" in politics in Paul Saurette, "You Dissin' Me? Humiliation and Post 9/11 Global Politics," *Review of International Studies* 32 (2006): 495–522.

30. Rana Mitter, *A Bitter Revolution: China's Struggle with the Modern World* (Oxford: Oxford University Press, 2005), 119–23.

31. Jungmin Seo explained that Chinese nationalism constitutes a "hegemonic politico-social discourse" which has undergone very few challenges since May Fourth. See Jungmin Seo, "Nationalism and the Problem of Political Legitimacy in China," in *Legitimacy: Ambiguities of Political Success or Failure in East and Southeast Asia*, ed. Lynn White (Singapore: World Scientific, 2005), 154–78.

32. John Fitzgerald, "The Nationless State: The Search for a Nation in Modern Chinese Nationalism," in *Chinese Nationalism*, ed. Jonathan Unger (New York: M. E. Sharp, 1996), 57.

33. John W. Garver, *Foreign Policy of the People's Republic of China* (Englewood Cliffs, N.J.: Prentice Hall, 1993), 20.

34. Anthony Giddens, *The Constitution of Society: Outline of the Theory of Structuration* (Cambridge: Polity Press, 1984); also John B. Thompson, "The Theory of Structuration," in *Social Theories of Modern Societies: Anthony Giddens and his Critics*, ed. David Held and John B. Thompson (Cambridge: Cambridge University Press, 1989).

35. C. P. Fitzgerald, *Revolution in China* (London: Cresset Press, 1952), chapters 2 and 3; Hughes, *Taiwan and Chinese Nationalism*, 7–16; Chalmers A. Johnson, *Peasant Nationalism and Communist Power: The Emergence of Revolutionary China 1937–1945* (Stanford, Calif.: Stanford University Press, 1962); also Jian Zhang, "The Influence of Chinese Nationalism on Sino-Japanese Relations," in *China-Japan Relations in the Twenty-First Century: Creating a Future Past?* ed. Michael Heazle and Nick Knight (Cheltenham: Edward Elgar, 2007), 23; Gunter Schubert, "Nationalism and National Identity in Contemporary China: Assessing the Debate," *Issues and Studies* 37 (2001): 134–37; Frederick C. Teiwes, "The Establishment and Consolidation of the New Regime, 1949–57," and Kenneth Leiberthal, "The Great Leap Forward and the Split in the Yan'an Leadership, 1958–65," both in *The Politics of China: The Eras of Mao and Deng*, second ed., ed. Roderick MacFarquhar (Cambridge: Cambridge University Press, 1997); Suisheng Zhao, "Chinese Nationalism and Its International Orientations," *Political Science Quarterly* 115 (2000): 3–10.

36. Hughes, *Taiwan and Chinese Nationalism*, chapters 1, 2, and 4.

37. CNNIC, *Statistical Survey Report*, 14–37.

38. See, *inter alia*, Thomas A. Metzger and Ramon H. Myers, "Chinese Nationalism and American Policy," *Orbis* 42 (1998): 21–36; Thomas Christensen, "Chinese Realpolitik: Reading Beijing's World View," *Foreign Affairs* 75 (1996): 37–52; Zhao, "Chinese Nationalism and Its International Orientations," 1–33; Alan R. Kluver, *Legitimating the Chinese Economic Reforms: A Rhetoric of Myth and Orthodoxy* (Albany, N.Y.: SUNY Press, 1996); Kalpana Misra, *From Post-Maoism to Post-Marxism: The Erosion of Official Ideology in Deng's China* (New York: Routledge, 1998).

39. Zhao, "Chinese Nationalism and Its International Orientations," 9.

40. Suisheng Zhao, "A State-Led Nationalism: The Patriotic Education Campaign in Post-Tiananmen China," *Communist and Post-Communist Studies* 31 (1998): 289; also Suisheng Zhao, *Nation-State by Construction: Dynamics of Modern Chinese Nationalism* (Stanford, Calif.: Stanford University Press, 2004), chapter 6.

41. The author wishes to thank an anonymous contributor for this point made during an interview in Singapore, 10 February 2008.

42. Zhao, "A State-Led Nationalism," 293.

43. Suisheng Zhao, "Chinese Intellectuals' Quest for National Greatness and Nationalistic Writing in the 1990s," *The China Quarterly* 152 (1997): 725–745.

44. Qiang Song, Zangzang Zhang, Ben Qiao, Qingsheng Gu, Zhengyu Tang, *China Can Say No* (*Zhongguo keyi shuo bu*) (Beijing: Zhonghua gongshang lianhe chubanshe,

1996)—this is the first of the series of "Say No" literatures in China in the mid-1990s. Geremie R. Barmé, "To Screw Foreigners is Patriotic: China's Avant-Garde Nationalists," in *Chinese Nationalism*, ed. Jonathan Unger (New York: M. E. Sharpe, 1996).

45. From a literary perspective, there are many hidden notes in the series that suggest such a love-hate relationship with the state (or regime, as conflated in the books). While the authors criticized the current leadership on their inferiority complex vis-à-vis the United States, arguably their dissent had originated more out of their loathing of the United States than of the leadership, and there was constant mentioning of the "glory days" of the old Communist leadership that were willing to stand up for the nation. To cite just one example, consider these words:

> In the 1940s, when American journalist Agnes Smedley visited Yan'an, she had used her female instincts to discover: that in the hearts of Mao Zedong and his generation of Red Army chiefs, there was this thick door that was tightly closed. This was a door protected by [China's] 5,000 years of civilization, the Great Wall and the Forbidden City, and was a door coated with layers of uprising and reforms. . . . And when this door opened, it was the moment in 1949 when the People's Republic of China was established. At that very moment, the door opened with a loud roar; and at that very moment, it sent out a strong and powerful voice: the Chinese nation which represents a quarter of whole mankind from thereon has stood up!

From a sceptical perspective, the underlying trusts on the regime and the State were perhaps the main reasons why the series was eventually "rehabilitated" after a ban was initially passed against its publication in mainland China. For words translated by this author, see Qiang Song, Zangzang Zhang, Ben Qiao, Qingsheng Gu, Zhengyu Tang, Preface to *China Can Still Say No* (*Zhongguo haishi neng shuo bu*) (Beijing: Zhongguo wenlian chubanshe, 1996), ii.

46. China was first recognized to be an Internet accessible country in 1994; until 1997, when the first CNNIC report was compiled, there were only an estimated 299,000 computer hosts and 620,000 Internet users. Access to international news and media was limited. Even today, major international news websites remained blocked by the firewall controls managed by the Public Information Network Security Supervision Bureau, a subsidiary of the Ministry of Public Security. See China Internet Network Information Center, *Statistical Survey Report on the Internet Development in China No. 1* (Beijing: CNNIC, October 1997), 1; also see China Internet Network Information Center, "The Internet Timeline of China" [www.cnnic.cn/en/index/0O/06/index.htm, accessed 10 August 2008].

47. Dingxin Zhao, "An Angle on Nationalism in China Today: Attitudes among Beijing Students after Belgrade 1999," *The China Quarterly* 172 (2002): 887.

48. "Schools Must Give Priority to Correct Political Orientation, Says Li Peng," *Xinhua News Agency*, 21 March 1990; "China Says 'Study Opium War' as Bulwark Against Imperialists," *Reuters*, 31 March 1990; "Symposium Marking Opium War Held in Beijing," *Xinhua News Agency*, 3 June 1990.

49. "Beijing Says Olympics Would Show Chinese Rebirth," *Reuters*, 5 September 1993.

50. Opinions differ widely regarding the impacts of, for example, Samuel Huntington's "clash of civilizations" theory, and comparable theses like Robert Bernstein's and Ross Munro's *The Coming Conflict with China*, in the Chinese thinking. But according to Dinxin Zhao's observations, such provocative publications are likely to capture more general attention than "more serious writings." Samuel P. Huntington, *The Clash of Civilizations and the Remaking of World Order* (New York: Simon & Schuster, 1996); Robert Bernstein and Ross H. Munro, *The Coming Conflict with China* (New York: Knopf, 1997); also Zhao, "An Angle on Nationalism in China Today," 902–3.

51. Zhang, "The Influence of Chinese Nationalism on Sino-Japanese Relations," 30.

52. Meixin Pei, "The Dark Side of China's Rise," *Foreign Policy* 153 (2006): 34.

53. Central Intelligence Agency, *The World Factbook* [www.cia.gov/library/publications/the-world-factbook/, accessed 1 August 2008].

54. In his 1997 New Year speech broadcast over the state media, Deng's successor Jiang Zemin described the retrocession of Hong Kong as "using snow to wash up the blood of the hundred years of national humiliation" (*xuexi bainian chiru*). *Xuexi* is a Chinese traditional aphorism that conveys an emotional redress over a grave misdeed and an exoneration of self-guilt and shame. See Jiang Zemin, "Working Together to Build a Better Future (transcript)," TRS WAS [www.baic.gov.cn/was40/detail?record=1&channelid=5877&searchword=%E7%89%88%E6%AC%A1%3D1, accessed 12 August 2008].

55. "China's Central Sports Bureau Chief Visits Taiwan; Yuan Weimin: Bidding for the Olympics is the Hope of all Chinese around the World," *Lianhe Zaobao*, 9 May 2001; "Beijing in Ecstasy after Successful Olympic Bid," *Lianhe Zaobao*, 14 July 2001.

56. Although observers have noted that on the eve of China's entry into the WTO, Beijing tried to keep unexpectedly low profile, the nation's expectations were pushed to a height after continuous reminders about the "arduous" fifteen-year negotiation process in the lead-up to the actual day of WTO entry. See "China Enters the WTO in Low Profile," *Lianhe Zaobao*, 12 December 2001.

57. "Liberating the Spirit of being Chinese," *Lianhe Zaobao*, 16 October 2003.

58. Yongnian Zheng and Sow Keat Tok, "'Harmonious Society' and 'Harmonious World': China's Policy Discourse under Hu Jintao," *Briefing Series* 26 (Nottingham: China Policy Institute, 2007), 7–10.

59. Chengju Huang, "The Development of a Semi-Independent Press in Post-Mao China: An Overview and a Case Study of *Chengdu Business News*," *Journalism Studies* 1 (2000): 649–64.

60. Wei Zhou (ed.), *Report on the Prospects of Chinese Media: Changes to an Industry and its Future Developments* [*Meiti qianyan baogao: yige hangye de biange quanjing he weilai zouxiang*] (Beijing: Guangming chubanshe, 2002), chapter 1.

61. Zhou, *Report on the Prospects of Chinese Media*, 20–51.

62. Huang Chengju, "Trace the Stones in Crossing the River: Media Structural Changes in Post-WTO China," *The International Communication Gazette* 69 (2007): 417–19.

63. Susan Shirk "Changing Media, Changing Foreign Policy in China," *Japanese Journal of Political Science* 8 (2007): 47.

64. For an example, see Xu Wu, *Chinese Cyber Nationalism: Evolution, Characteristics, and Implications* (Lanham, Md.: Lexington Books, 2007).

65. Yong Deng, "Hegemon on the Offensive: Chinese Perspectives on U.S. Global Strategy," *Political Science Quarterly* 116 (2001): 349–52. For other examples, see various articles in the special online collection about the 1999 Belgrade bombing, "First Anniversary of the Bombing of Chinese Embassy in Yugoslavia," Sina.com, news.sina.com.cn/weekend/kosovo.shtml [accessed 15 November 2008]; also Jiang Zemin's linking "new interventionism" to "new hegemonism" in "China Urged Caution toward New Interventionism" (*zhongguo cu jingti xin ganshe zhuyi*), *Wen Wei Po*, 26 August 1999.

66. See, for an example in the continuing volatility of anti-U.S. sentiments in China, especially via the Internet, "Assault on Sino-US relations," *South China Morning Post*, 17 August 2004; also "Government Faces Intellectuals' Wrath for 'Kowtowing' to US," *Straits Times*, 17 February 2003.

67. Lam Peng Er, "Japan's Deteriorating Ties with China: The Koizumi Factor," *China: An International Journal* 3 (2005): 284–85.

68. Sow Keat Tok, "Neither Friends nor Foes: China's Dilemmas in Managing its Japan Policy," *China: An International Journal* 3 (2005): 292–300.

69. While Beijing has imposed domestic media clampdowns on reports over China's submarine incursions into the Japanese territorial waters in the past years, articles reporting on these events nonetheless seeped through the censors and were republished on private and commercial sites, Internet chatrooms and personal blogs. These reports often capture the attention of ardent online Chinese nationalists. See, for example, a personal blog, by nationalist blogger "Philosophy in Poems" (*shizhong de zhexue*) on the popular website *Sohu*, which described one of such incursions as an "elation following long periods of oppression" (*yangmei tuqi*). Similarly, other reports also described Japanese "embarrassments" over its failure to stop such incursions, rather than vice-versa. See Philosophy in Poems Blog, "Thoughts on the Day of National Shame" (*guochiri zhi gan*) (18 September 2008), drshizheng.blog.sohu.com/100120107.html; "Japan Extremely Embarrassed over 'Unidentified Submarine' Penetration into its Anti-Submarine Network" (*buming qianting churu riben fanqianwang; riben hen ganga*), *Northeast Military News*, 19 September 2008, mil.nen.com.cn/military/195/3171195.shtml [both accessed 12 November 2008].

70. See for example, the blog article and subsequent comments in Tang Ti Blog, "The Focus of Chen Yunlin's Visit" (*chen yunlin fang tai de jiaodian*) (3 November 2008), blog.sina.com.cn/s/blog_53a53ebe0100b3j7.html?tj=1 [accessed 19 November 2008]. Meanwhile, China's official forums like "State Strengthening Forum" [*Qiangguo luntan*] also recorded a lot of positive responses on the issue.

71. See, in particular, chapters 1(4), 3(26), 6(64), 9(91), and 11(109). "The Constitution of the Republic of China," Government Information Office, [info.gio.gov.tw/ct.asp?xItem=13370&CtNode=904&mp=1, accessed 12 August 2008].

72. "Four Types of Nationalisms in the Fight over Mongolia," *Ming Pao*, 19 December 2004.

73. "China, Russia Sign Border Agreement: Foreign Ministers," *Agence France Presse*, 21 July 2008; "Leaders Sing Praises of Sino-Indian Teamwork," *South China*

Morning Post, 15 January 2008; "Wen Urges Understanding, Accommodation in Addressing Disputes," *Xinhua News Agency*, 7 October 2004.

74. Wenhua Shan, "Redefining the Chinese Concept of Sovereignty," *China and the New International Order*, ed. Wang Gungwu and Zheng Yongnian (London: Routledge, 2008), 70–72.

75. Spence, *The Search for Modern China*, 237–38, 288–94, 746–47.

76. This is in contrast to those already pessimistic views offered by other academics. See, for example, Peter Hays Gries, "Chinese Nationalism: Challenging the State?" *Current History* 104 (2005): 251–56.

77. For an account of the events, see Tok, "Neither Friends nor Foes: China's Dilemmas in Managing its Japan Policy," 292–300; "China Blocks Anti-Japan Websites," *Sydney Morning Herald*, 3 May 2005.

78. Zhang, "The Influence of Chinese Nationalism on Sino-Japanese Relations," 24–25; Gries, "Chinese Nationalism," 252; "Japan Military Transport to China; Beijing Says More Consultation Needed," *Xin Pao*, 30 May 2008.

79. Tamara Renee Shie, "The Tangled Web: Does the Internet Offer Promise or Peril for the Chinese Communist Party?" *Journal of Contemporary China* 13 (2004): 537–38.

80. For an interesting exploration of the concept of "enduring injustice," see Jeff Spinner-Halev, "From Historical to Enduring Injustice," *Political Theory* 35 (2007): 574–97.

81. Liang Fook Lye, "The Party and the Media: Control versus Change," in *Interpreting China's Development*, ed. Wang Gungwu and John Wong (Singapore: World Scientific, 2007), 25–27.

82. See Shie, "The Tangled Web," 529–36.

II

THE FIRST LAYER: JAPAN, TAIWAN, AND THE UNITED STATES

3

China's Online Nationalism Toward Japan

James Reilly

Does the Internet in China contribute to liberalism or nationalism? Studies that focus on the Chinese state's capacity to control the flow of information online generally presume that widespread, unfettered Internet access would encourage political opposition to the Communist Party, greater intellectual diversity, and a more vibrant civil society.[1] Indeed, a number of studies suggest that growth of the Internet in China has strengthened societal oversight of the state and empowered civil society actors such as environmental groups.[2] The case of foreign policy may, however, be more complex. Even if the Internet does strengthen civil society autonomy, deliver more information to a wider public, and build horizontal linkages among civil society groups, such trends may not move Chinese foreign policy in a more liberal or pragmatic direction.

In the case of China's Japan policy from 2002 to 2005, widespread use of the Internet contributed to a wave of popular nationalism which undermined pragmatic and liberal approaches to Japan. Online popular nationalism exacerbated negative public sentiments toward Japan, impeded bilateral economic cooperation, and undermined liberal and pragmatic discourses over foreign policy. After Chinese authorities began restricting online activism and popular media coverage in 2005, foreign policy toward Japan began to shift and public opinion toward Japan improved. This case suggests that in an authoritarian state like China, the Internet is more likely to foster nationalism than liberalism, while restraints imposed by the authoritarian state may contribute to a more pragmatic foreign policy.

This chapter begins with an overview of China's popular nationalism toward Japan, detailing the role of Internet activists, Internet chatrooms, and news websites. The subsequent three sections describe the impact of online nationalism on public opinion, on Japanese involvement in building a high-speed rail line in China, and in online debates over the "new thinking" toward Japan. The conclusion explores implications of these findings for the impact of the Internet on China's overall foreign policy.

China's Online Nationalism Toward Japan

China's "year of Internet nationalism" toward Japan began in July 2003 with the "Alliance of Patriots" website's online petition campaign in opposition to Japanese companies' involvement in building a high-speed rail line (*shinkansen*) in China, which netted over eighty thousand signatures in ten days.[3] The following month, a petition campaign in protest of the "Qiqihair incident" in which abandoned Japanese chemical weapons were accidentally unearthed, leaving one person dead and thirty-seven severely injured, collected over a million signatures in one month via 12,518 websites.[4] Online emotions grew stronger following reports of a sex orgy in Zhuhai by Japanese businessmen on September 18, student riots in Xi'an in response to a licentious skit by Japanese university students on October 29, and riots after a China-Japan soccer match in August 2004. Online controversies also emerged over Japanese car advertisements, Japanese video games, and treatment of Chinese passengers on Japanese airlines. In spring 2005, over ten million people signed online petitions opposing Japan's becoming a permanent member of the UN Security Council.[5]

China's online nationalists describe themselves as enacting the public's right, even its obligation, to defend China's national interests by challenging external threats and stiffening the government's resolve in defending these interests. Contrary to Suisheng Zhao's claims, they are populists but not liberals.[6] They do not advocate universal liberal ideals, nor do they pursue a more liberal or democratic system of governance. Instead, they prioritize a militarily and economically powerful Chinese state, capable of defending all of China's territorial claims while demanding international respect. They tend to adopt a zero-sum perspective of economic and security relations with Japan, and do not trust ideational or institutional constraints upon Japanese power. Popular nationalism is fueled by emotional, negative sentiments toward Japan and Japanese people, and is engaged with ethnic Chinese worldwide.[7]

China's online nationalists tend to move from issue to issue, reacting spontaneously to recent developments in China-Japan relations. Cognizant of the

CCP's fear of organized opposition, activists are careful to insist that they are all "individual volunteers, not supported by any organization, and not taking any extremist actions."[8] They are praised by journalists as "part of the Chinese masses; not some intellectuals' salon."[9] One reason is their emotional challenge to the state's emphasis on remaining "calm and reasonable" (*lengjing yu lizhi*). As Feng Jihua, a leading activist, stated in a 2004 interview:

> As a Chinese person, when I see the history written there [at the history museum attached to Yasukuni], the people they respect and the spirits they promote, I cannot be "calm and reasonable." It just makes a person "excited and furious." If we could do some objective introductions of Yasukuni Shrine for the average Chinese reader, I believe they too would feel furious.[10]

Activists' Anti-Japan Websites

The first activist website on Japan-related issues, "China918," was established by Wu Zukang, a middle-aged engineer, and his daughter in March 2000 to repudiate right-wing Japanese groups' accounts of the Nanjing Massacre.[11] In 2001, website organizers held a series of activities to commemorate the seventieth anniversary of the September 18 (Mukden) incident. Television reports of the activities caught the attention of overseas Chinese, who began to support the website financially. In 2002, the website issued a call for an "online war of resistance" against Japanese right-wing and revisionist groups, inspired by the actions of Feng Jihua.[12]

Feng Jihua was a Chinese worker living in Japan who came to fame on August 14, 2001 (the day after Prime Minister Koizumi's first visit to Yasukuni), when he spray painted *gaisi* ("go to hell") on a wall of the Yasukuni temple.[13] After his widely published court statement in Japan, entitled "Chinese people's right to resist," Feng Jihua was selected as a "Person of the Year" by readers' polls by *Southern Weekend* and Sina.com in 2001, with many praising him as "China's backbone."[14] Upon his return to China on June 23, 2002, he was invited by Lu Yunfei to join the newly founded "Alliance of Patriots" website.[15] By spearheading the *shinkansen* and Qiqihair online petition campaigns, Lu Yunfei and Feng Jihua catapulted their website to the forefront of China's online activism, becoming influential media figures in China and abroad.[16] The website quickly spawned numerous followers: by 2004, China was host to some nine thousand Japan-focused activist web pages.[17]

As of 2006, the Alliance of Patriots website was run by seventy volunteers with over 130,000 registered members.[18] They are overwhelmingly young and male.[19] The website is divided into nineteen geographical regions, each of which is headed by a locally based coordinator. Each location included a chat room where "web friends" could get to know each other online, and then meet in person. The website also includes twenty-four issue-based discussion "rooms," which facilitate online and off-line coordination among activists. In 2007, the website established one of China's first e-magazines entitled "China's Diaoyu Islands," designed to facilitate civil society discussions and coordinate activism.[20]

Online Chatrooms and News Websites in China

Activist websites' impact was greatly augmented by the rapid growth of China's active online community. By the end of 2007, China's "netizens" (regular Internet users) numbered 210 million, a 53 percent increase over 2006.[21] Chinese netizens tend to be young urban residents, with above-average education and income levels.[22] A nationwide poll in 2007 found that 64 percent of university students use the Internet as their primary source of information on Japan, a trend that has been rising in recent years.[23] A CASS poll in October 2004 found that a quarter of all respondents from sixteen- to fifty-nine-year-olds rely primarily on the Internet for their information on Japan.[24] Over 60 percent of Chinese "societal elites" rely primarily on the Internet for international news, as opposed to only 6 percent of the general public.[25] Another poll found that over half of China's middle class use the Internet as one of their main sources of news.[26] A 2007 survey found that almost 20 percent of Chinese netizens' first stop online is to check news websites.[27] For-profit websites such as Sina.com and Sohu.com dominate in providing news to the general public; state-run media sites such as Xinhua's website are far less popular.[28]

Chinese online discussions are carried on in three venues: specialized "bulletin board" sites (BBS), discussion forums maintained by Chinese news websites, and forums maintained by websites such as the "Patriots Alliance." These online forums allow Chinese netizens to easily get the latest news, express their opinions publicly, and engage in political activism online. For instance, during the spring 2005 Japan protests, the popular Sina.com website's top stories contained links to related news coverage, online petitions, and specialized bulletin boards.

Two of the most important BBS sites for Japan-focused online nationalism are maintained by the *People's Daily* online (*renminwang*): the "Strong Nation Forum" (*Qiangguo luntan*) and the "China-Japan Forum" (*zhong-ri luntan*). As of April 19, 2008, the Strong Nation Forum had 623,092 registered users, with an average of some three to four million users online at any one time.[29] The China-Japan Forum, established on January 18, 2000, hosts lively policy debates on China-Japan relations.[30] For instance, a Chinese expert's 2005 article in *Global Times* questioning if "Japan will become Asia's England" garnered twelve thousand "hits" in only one day after being posted on the China-Japan Forum.[31] The Sina.com website claims to host the largest and most active Chinese language Internet community in the world, home to over one hundred thousand postings every day. During the 2004 Olympic Games in Athens, 150 million people viewed Sina.com stories; "hits" on articles on the Carrefour boycott in spring 2008 reached over half a million.[32] The Tianya.com neighborhood (*shequ*) declares itself the "online home of Chinese people worldwide": as of April 19, 2008, Tianya had approximately twenty thousand official "members," with approximately ten thousand online at any one time.[33]

From 2002 through 2005, the growing influence of news websites and online chat rooms combined with China's market-oriented media to spread anti-Japan sentiments among the public. Although news websites can only reprint stories which have already appeared in legitimate print media, they found ready fodder among the sensationalist, often negative, coverage of Japan in China's most influential, cutting-edge newspapers. One study of *Huanqiu Shibao* (Global Times) documents a dramatic rise in coverage of Japan and history-related issues from 2004 to 2005, exacerbating anti-Japanese "parochial nationalism" (*xia'ai minzhuzhuyi*).[34] Sensationalist stories of Japan-related incidents in obscure city-level "evening papers" (*wanbao*), less regulated due to their traditional focus on "soft news" such as entertainment, culture, and local events, were also spread nationwide by online news sites. Mainstream media also publicized online activism: activist Tong Zeng's open letter to UN Secretary-General Kofi Anan in spring 2005 was initially posted online and then published in full by a major weekly newsmagazine.[35]

One example of this media-Internet multiplier effect was the uproar over a photograph of Chinese model/actress Zhao Wei in the September 2001 issue of the popular magazine *Shizhuang* (Fashion) wearing a dress imprinted with a large Japanese imperial flag. There was no noticeable public reaction in China until a local Hunan newspaper ran a story on the incident in December 2001. The article was widely reprinted by Internet news sites and then covered by many other urban newspapers around China. The subsequent uproar on the Internet forced the editor of *Shizhuang* to resign. Zhao Wei's house was smashed and she was personally attacked in public, and later forced to make a public apology, broadcast on national television.[36]

Popular media and the Internet also combined to spread public anger and compel an official response to a three-day sex scandal involving Japanese businessmen in Zhuhai, China from September 15 to September 18, 2003. There were no media reports or official responses to the incident until *China Youth Daily* ran an indignant report on September 26.[37] By that afternoon, the *China Youth Daily* story was on the front page of all major Chinese news websites, was reprinted in a number of urban newspapers' afternoon editions, and had become the hottest topic in Chinese online chat rooms. Local authorities immediately began investigating the case, prompting a second *China Youth Daily* story.[38] Online emotions ran high. In a Soho.com online poll on the incident, 90 percent of respondents agreed that the Japanese involved intended to humiliate China by picking September 18 as the date.[39]

The Internet also provided an opportunity for quick, widespread responses to such sensationalist news reports on Japan. Within five days after the Zhuhai prostitution incident was reported, over twenty-four thousand comments were left on the Sina.com website. Within one month after the March 24, 2004, arrest of Chinese *baodiao* activists by Japan, over 1.23 million responses were posted on Sina.com.[40] In an April 2005 online poll on Japan's history textbooks, eight million "votes" were registered on Sina.com within two days.[41] A 2003 online petition campaign calling for a boycott of Japan's Asahi Beer to protest its reputed support for Japanese right-wing groups garnered 1.28 million signatures in only two hours.[42] Such rapid online responses present a challenge for state censors. When a British website posted pictures of the Chinese vice-foreign minister appearing to bow to his Japanese counterpart after the spring 2005 protests, the site had 450,000 hits in only three and a half hours, before the Chinese government learned of the picture and blocked access to the site from China.[43]

One reason that sensationalist media coverage of Japan sparked such wildfires of anger and activism is the high level of participation in online chat rooms by Chinese netizens. A 2004 survey found that 21 percent of China's middle class participate in online discussions on a regular basis.[44] A 2002 survey of Chinese youth in six cities found that 23 percent "regularly" issued online comments, while 58 percent reported doing so "occasionally."[45] As a leading Chinese expert explained in 2004, "The Internet has become the most important space for the Chinese public to express its opinion."[46]

In reaction to this wave of online nationalism in 2003, a few Chinese scholars, journalists, and officials began to challenge China's "angry youth" (*fenqingmen*). A PhD student at Beijing University criticized the *shinkansen* petition campaign, arguing that it was "undemocratic" to value the opinion of eighty thousand people over the interests of 1.3 billion.[47] Scholars denounced online protests as "parochial nationalism," while others warned that such "popular diplomacy" (*minjian waijiao*) would upset delicate diplomatic negotiations.[48] Yet most postings challenging popular nationalism soon found themselves a target for virulent criticism. As Liu Xiaobiao, an influential journalist and scholar, explained in January 2004: "If the topic concerns Japan, the great majority of views expressed will be criticism, condemnation, and even abuse in fierce terms. Anyone who puts forward a critical or moderate voice challenging extreme nationalism is strongly denounced, resulting in a 'screw of silence.'" He noted that China's Japan experts had begun requesting journalists not to reveal their names, so as to avoid being attacked as a "traitor" online. The result, he warned, was that online nationalism toward Japan was "hijacking" Chinese patriotism.[49] To investigate this claim further, the following three sections evaluate the impact of online nationalism on public opinion, policy-making, and policy discourse.

Online Nationalism and Public Opinion Toward Japan

The rise in online nationalism was supported by, and contributed to, a worsening of public opinion toward Japan from 2002 to 2006. For instance, three polls conducted by Institute of Japan Studies (IJS) of the Chinese Academy of Social Science (CASS) show that Chinese feelings of closeness to Japan declined (see figure 3.1 below).

From 2002 through 2006, Chinese attitudes toward Japan became more polarized and strongly negative. The largest shift of any category was an increase in those feeling "not close" to Japan, up from 27.7 percent in 2002 to 31.8 percent in 2006; and "very not close," which went up from 15.6 percent in 2002 to 23.4 percent in 2006. Those who were ambivalent toward Japan declined from 47.6 percent in 2002 to 35.5 percent in 2004. The impression of a more polarized Chinese public is reinforced by a comparison between the 2002 CASS survey and a 2005 survey by Beijing University, in which negative impressions of Japan went up 19 percent, while "average" impressions

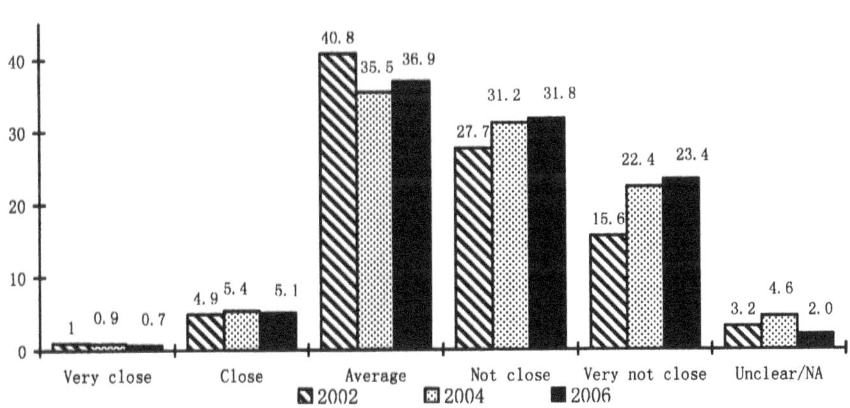

FIGURE 3.1
2002–2006 Feelings of Closeness to Japan
Source: Institute of Japan Studies of CASS, surveys 2002–2006[50]

of Japan dropped 20 percent.[51] Other trend indicators also suggest a growing polarization of attitudes toward Japan from 2002 to 2005. In the 2002 CASS poll, 66.5 percent felt no or little "affinity to Japan" (28.2 percent and 38.3 percent respectively). Three years later, in 2005, 71.1 percent held this view (24.6 percent and 46.5 percent respectively). When asked in October 2004 if they "liked" Japan, 38.8 percent responded no. In June 2005, the negative view increased to 46.0 percent.[52] This trend is further reinforced by a poll conducted in May and June 2005, in which 62.9 percent reported negative impressions of Japan.[53]

The decline in public opinion toward Japan seems to have been caused more by the rise in sensationalist, negative coverage by market-oriented media and its distribution online, rather than by state propaganda. For instance, the Chinese government steadily increased its criticism of Prime Minister Koizumi's annual visits to the Yasukuni Shrine, particularly after the fourth visit on January 1, 2004. If state propaganda was the primary factor causing a worsening of public opinion, we would expect that the public held ever-stronger attitudes toward Yasukuni during this time period. However, the three CASS polls show that Chinese public opinion on Yasukuni visits was actually strongest in 2002, and slightly moderated over the next four years. The number of respondents willing to accept prime ministerial visits to Yasukuni under certain conditions rose from 2002 to 2006, while those who felt that under no conditions should Japanese leaders go to Yasukuni was highest in 2002, with a significant drop off in 2004 and 2006.[54] Furthermore, the rate of Chinese people who associated Japan with the image of Yasukuni Shrine was initially quite high in 2002, declined in 2004, and then rose again in 2006.[55]

Although Yasukuni did not increase in importance to the Chinese public over this period, history issues in general did. The CASS surveys asked respondent who did not feel close to Japan to list their reasons. The two most common choices were "Japan's invasion of China" or "because Japan to this day has not thoroughly 'reflected on' (*fanshen*) its invasion of China." The percentage of these responses rose steadily from 2002 to 2006. At the same time, those who answered that they felt not close "because the US-Japan military alliance is a threat to China's security" declined (see figure 3.2). Echoing this trend, the percentage of people who listed history issues as the most important factor for the healthy and stable development of China-Japan relations in the CASS polls rose 10.4 percent from 2004 to 2006, overtaking economics and trade.

Like the Yasukuni issue, the Chinese government put strong emphasis upon criticizing the U.S.-Japan security alliance during the 2001–2006 period, particularly in 2005 when the alliance framework was strengthened and

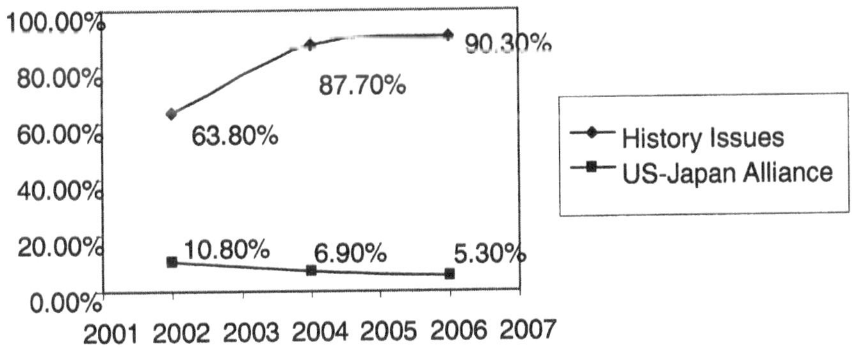

FIGURE 3.2
Reasons For Not Feeling Close to Japan: 2002–2006
Source: Institute of Japan Studies of CASS, surveys 2002–2006[56]

referred indirectly to Taiwan. The steady decline in public concern with the alliance thus does not seem to indicate a strong impact of government propaganda upon public opinion. In an authoritarian state like China, it is difficult to assess the relative impact of the state propaganda versus the impact of the Internet and market-oriented media on public opinion. Yet the evidence suggests that sensationalist, negative coverage of Japan-related issues distributed online played a key role in exacerbating negative opinions of Japan among the Chinese public.

Impact of Online Nationalism on Foreign Policy: The *Shinkansen* Case

Contrary to common presumptions, online nationalism in authoritarian states like China may even have a greater impact on foreign policy than in democracies. Unable to credibly claim that their authority derives from a meaningful public choice, authoritarian leaders tend to rely more upon nationalist rhetoric than democratic leaders to sustain popular support and justify their rule. Broadly based nationalist protests appeal to the state's legitimating nationalist rhetoric, and so ignoring or publicly suppressing these protests is particularly risky for leaders. In addition, the scarcity of legitimate channels for authoritarian publics to express their preferences means that although public opinion is more constrained, expressions of nationalism tend to be more explosive and extreme. In foreign policy, issues of nationalism appeal strongly to both state legitimacy and popular passion.[57] As such, authoritarian leaders must be particularly cautious that popular protests on nationalist issues do not grow quickly out of hand, undermining domestic stability and challenging leaders' legitimating ideology.

To navigate these treacherous straits, Chinese leaders expend great resources monitoring public sentiments, Internet debates, and popular media coverage on nationalist issues. For instance, the Chinese Propaganda Ministry provides top leaders with a daily short report (*jianbao*) on media reporting of contentious foreign policy-related issues. Each ministry also produces a daily "Internet report" (*hulianwang jianbao*); the reports compiled by Xinhua News Agency and the Public Security Bureau are reported to be particularly significant for policy-making.[58] As Alistair Iain Johnston notes:

> There is considerable evidence that the regime focuses on more extremist attitudes (as found on the Internet bulletin boards, for instance), presumably because these views are a barometer of the kinds of emotions that would get protesters into the streets. And it is street protests on issues such as Taiwan or the United States—linked to intra-elite competition—that the current leaders appear to worry about.[59]

Indeed, in the first of his series of online chats on December 23, 2003, Chinese Foreign Minister Li Zhaoxing told Chinese netizens: "I've read online comments before that our diplomacy is sometimes too 'soft' (*ruan*) or too 'hard' (*ying*). We take such concerns seriously."[60]

In the face of mounting criticism, Chinese policy-makers may find it less costly to appease the public's demands by altering their foreign policy choices than to either repress or ignore public pressure. Online nationalism can thus affect the timing and direction of foreign policy decisions, as well as negotiating behavior and official rhetoric. Online protests can affect the content and scope of negotiators' win-sets, creating an interest in achieving a certain outcome while excluding other outcomes from officials' acceptable win-sets. In the case of China's decision to build a high-speed rail line (*shinkansen*), online nationalism broadened the scope of the decision-making process, delayed decisions, interjected the element of public emotion into officials' calculations, and so affected the final decision.

Building a High-Speed Rail Line in China

After years of debate, in 2002 Chinese officials began plans to issue contracts to foreign firms to help construct a high-speed train linking Beijing and Shanghai. The overall project was estimated in 2003 to be worth over US$10 billion, and offered foreign firms a potential monopoly on what might quickly become the world's largest high-speed train market.[61] While a German consortium was initially in the lead due to its involvement in the Shanghai railway line, in 2002 disputes arose over technology transfer, cost overruns, and technical problems.[62]

Around the same time, Chinese officials openly warmed to the Japanese bid. On September 24, 2002, Chinese leaders meeting with Chikage Ogi, Japan's Minister of Transportation, suggested that perhaps *shinkansen*-type technology would be more appropriate for China instead of the German maglev. Ogi immediately returned to Japan and put together a high-powered, civil "Japan *Shinkansen* Promotion Committee" headed by former Prime Minister Fukuda. This committee visited China several times over the next year to promote the project.[63] At the same time, a number of China's Japan experts urged the Chinese government to offer contracts to Japan, seeing the project as a chance to shift from "cold politics, hot economics" to "politics promotes economics, and economics promotes politics."[64]

In early 2003, a researcher familiar with the project from China's Railway Ministry's Science Research Institute stated that, "the possibility of selecting Japan's *shinkansen* was already greater than 90 percent. Even though it required approval by the top leadership, the essential explanation behind the approval was already done and the only thing left to do was the procedural aspects of the auditing."[65] The Railway Ministry cancelled its own high-speed rail project, and issued internal reports announcing the plan to award contracts to Japanese firms.[66] In April 2003, Japan's *Economic News* reported that "China's high level leaders have told Japanese railroad executives that the possibility of China choosing Japanese technology is 95 percent."[67] President Hu Jintao told Prime Minister Junichiro Koizumi on May 31, 2003, that he would like to factor China-Japan cooperation into the rail project.[68] Japanese officials expected that China would award contracts to Japan before the end of 2003, as part of commemorating the twenty-fifth anniversary of the Treaty of Peace and Friendship.[69] Before the summer of 2003, there was no sign of any significant debate over the domestic political ramifications of Japanese involvement.

Rise of Popular Nationalism Online

Chinese activists' involvement began with a comment posted on a Chinese online bulletin board in October 2002 warning that Japan was about to receive the high-speed rail contracts. Internet discussions grew more heated in spring 2003, as Japanese media reported that Japan was sure to gain contracts.[70] These reports finally spread to the Chinese-language press in mid-July, with similar reports in a Hong Kong newspaper.[71] In response, on July 19, 2003, the "Alliance of Patriots" website started an online petition by declaring "We refuse the use of Japanese products for the Beijing-Shanghai train."[72] They warned against allowing Japan to "reinvade" China, drawing parallels with the Manchurian railroad built by Japan and used in its invasion of China.[73] A number of other websites soon began to include links to this page.[74] On July 29, organizers submitted 82,752 signatures to the Ministry of Railroads in Beijing, declaring later to reporters that the petition "had a definite impact" on Chinese officials.[75] By early August 2003, a number of articles on the issue began to appear in influential Chinese newspapers and news magazines.[76] *Beijing Youth News* editorialized strongly in support of the sign-on campaign, arguing that although the Japanese government may have successfully convinced the Chinese government, it made a "strategic mistake" by not considering the emotions of the Chinese people.[77]

When Chikage Ogi, Japan's minister of transport, came to China on August 3, 2003, she found a radically different tone than just one year earlier. Zhang Guobao, the vice-director of the NDRC, the institution directly responsible for the project, told Ogi bluntly that "domestically in China, there is debate" over the Japanese technology, and explaining that "we will make a decision after listening to the Chinese citizens' public opinion."[78] Chinese railway officials later confirmed to reporters that China's cold reaction to Ogi's visit was due in large part to their concerns with public reactions.[79] In response, Beijing's *International Herald* declared that, "In the current situation, as Chinese citizens' rights of political participation are growing stronger, Internet nationalism is definitely having a clear influence upon government policy."[80] Other newspaper editorials praised activists for providing "support and reinforcement" for the Chinese government in standing up to Japan.[81]

On August 30, 2004, Chinese officials announced that Japanese firms had received contracts to build eighty high-speed trains, with a French consortium to supply sixty trains and Canadian firms to supply twenty trains.[82] Chinese activists quickly organized a new petition campaign urging "China Should Build China's Railways" (*Zhongguo tielu; zhongguo zao*), aimed at delivering one million signatures to the Railway Ministry.[83] The petition called for a reversal of the decision, as well as public hearings on the decision and the creation of a supervisory body under the National People's Congress to oversee the project. Within a single day, the Patriot Alliance collected 68,733 signatures online in support of the petition.[84] The next day, the website was shut down by government officials, and all online references to the petition campaign were blocked in China.[85]

Despite closing down the website, subsequent government decisions actually matched the demands of the protesters almost perfectly. The Chinese government pledged that it would use a more open bidding process to deal with subsequent bids.[86] Japanese companies failed to win any more contracts, even though German and French companies succeeded in doing so.[87] In December 2004, Vice-Foreign Minister Wu Dawei told a visiting Japanese Diet delegation, "We are facing high anti-Japan sentiment in China. If our government adopts the *shinkansen* technique in the railway project, the people would have [negative] opinions."[88] When foreign news agencies reported on November 23, 2005 that China was planning to order an additional sixty trains from the Japanese consortium, Chinese officials quickly and vociferously denied the claims in the Chinese media.[89] Finally, on March 7, 2006, the NDRC announced that China would build a conventional high-speed railway with Chinese technology for the Beijing-Shanghai line, and a mag-lev train between Shanghai and Hangzhou.[90] Apparently seeking to appease the public, Railway Minister Liu Zhijun announced, somewhat inaccurately, that the Beijing-Shanghai line would use entirely domestic technology.[91]

Impact of Online Nationalism

At first glance, China's decision may appear to be in reaction to Prime Minister Koizumi's repeated visits to the Yasukuni Shrine.[92] However, Chinese rhetoric, negotiating stances, and policy decisions more closely correlate with trends in public pressure. Chinese leaders repeatedly referred to public criticism on the rail project, but did not use the rail project as leverage to pressure Koizumi to not go to Yasukuni, though this would have been a plausible negotiating strategy. Up through June 2003, before online mobilization began, media reports in China and Japan, and public statements by Chinese leaders all indicated that Japanese firms were likely to be awarded rail contracts, even though Koizumi had already visited Yasukuni three times. Instead, the shift in policy came only after online mobilization picked up. There was also a discrepancy between policy on Yasukuni visits and the rail line. In November 2004, Chinese diplomats were quietly exploring possible solutions to the Yasukuni standoff with their Japanese counterparts, while at home Chinese officials were repeatedly denying that Japanese firms would win any further bids.[93] The final decision to use primarily Chinese firms was announced in March 2006, exactly at the point when Chinese leaders were already seeking ways to reengage the next Japanese prime minister.[94]

Online nationalism also shaped the decision-making process. Before spring 2003, the decision was dominated by a closed-door bureaucratic process controlled by the Railway Ministry and the NDRC in which technical and cost concerns were primary. After public protests began in summer 2003, the decision became a highly contested political choice with substantial domestic political costs for the Chinese leadership. Online protests interjected emotional factors into the decision process and forced a more public debate while delaying decisions. Railway officials later indicated in private interviews that public emotion became a critical factor in their considerations after 2003, even worrying that people might damage Japanese-built trains or otherwise sabotage the project if Japanese technology was used.[95]

The failure of the high-speed rail project, while an unmitigated success for China's online nationalists, represented a setback for both liberal and pragmatic approaches to Japan. A number of China's Japan experts had supported Japanese collaboration in the project as a mechanism to foster mutual trust and build common interests, similar to German-French economic collaboration after World War II.[96] For pragmatic Chinese leaders, the *shinkansen* protests marked a disturbing, if rare, instance of ideological sentiments prevailing over long-term strategic and economic preferences.

Impact of Online Nationalism on Public Discourse in China

The wave of anti-Japan online nationalism also affected public policy discourse in China, undermining liberal and pragmatic policy proposals. In the "new thinking" debate in late 2002 through early 2004, Chinese experts, public intellectuals, and the general public engaged in perhaps the most virulent and contentious round of public debate over foreign policy since 1949. Both liberal and pragmatic approaches were strongly denounced by nationalist, emotional responses online. The highly charged political atmosphere politicized policy discourse, rendering experts less willing to offer moderate policy proposals and making such moderation less attractive to policy-makers. Ultimately, the "new thinking" was only adopted by Chinese policy-makers and experts after the wave of online nationalism had subsided.

The new thinking debate began with a December 2002 article in *Zhanlue yu guanli* (Strategy and Management) by Ma Licheng, a member of the editorial board of the *People's Daily*. Ma first became an influential liberal voice in China after the publication of *Jiaofeng* (Crossing Swords), a book on the debates between reformers and leftists in the mid-1990s. After the publication of his second book in 1999, which criticized China's popular nationalism, Ma was selected as one of the fifty most influential people in China by *Asia Weekly*.[97]

Ma's 2002 article adopted an explicitly liberal approach. He criticized anti-Japanese nationalism in China as unhealthy, irrational, and destructive for China's own interests, urging China to take on "the attitudes of a victor and of a great power" and avoid "parochial nationalism" (*xia'ai minzuzhuyi*). Concerned that xenophobic sentiment in China was beginning to contain policy-makers, Ma urged that China "should not be too sensitive to Japan's quest to become a political and military power." Most controversially, he argued that

China had been too "harsh" (*keke*) toward Japan in demanding an apology for its wartime actions. Claiming that Japan had apologized twenty-one times, Ma urged that "the page on apologies should be turned; we should pass this period of history and face the future."[98]

In the second issue of *Strategy and Management* in 2003, Renmin University professor Shi Yinhong praised Ma's argument as "courageous," arguing that the increase in Sino-Japanese enmity is "quite dangerous." Shi advocated a pragmatic approach to Japan based upon the principle of "strategic centralization" (*zhanlue jizhong*). He argued that a rapprochement or "diplomatic revolution" with Japan would benefit China's strategic interests, enabling China to balance the pressure from U.S. hegemony and concentrate on the Taiwan issue. Shi also urged China to drop its pressure on history issues, strengthen bilateral economic ties, stop exaggerating the threat of Japanese militarism, and welcome Japan's full participation in regional and global politics as a "great power."[99] Shi later warned that in striving to base Japan policy upon strategy, not emotions, China's leadership faces the daunting but imperative challenge of "guiding and adjusting" (*yindao yu tiaokong*) popular opinion.[100] While Ma's approach was based upon liberal ideals, Shi Yinhong's argument was classic "realpolitik," echoing the enduring pragmatism in Chinese foreign policy. Neither approach was truly new: similar arguments had been broached publicly by Chinese experts repeatedly since 1998.[101]

After Ma and Shi's essays appeared, there was broad speculation that the essays represented a kind of "trial balloon" floated by the incoming Hu Jintao-Wen Jiabao leadership. Ma's article appeared just after the sixteenth Party Congress in which the new leaders signaled their interest in improving relations with Japan.[102] Ma Licheng was widely known to be close to the personal secretary of Zeng Qinghong, a member of the Politburo Standing Committee seen as moderate toward Japan. The *Yomuri Shinbum* reported that "a source close to the Chinese government said that Ma's essay came against the background of Zeng Qinghong's comment about thinking highly of Japan."[103] Both Ma and Shi later quietly indicated that they had several informal discussions with policy-makers both before and after their essays were published.[104]

Popular Nationalists Respond to the "New Thinking"

If the new thinking was indeed a trial balloon sent up by some members of the Chinese leadership, it was soon shot down by the onslaught of critical response facilitated by the Internet. The scope of the debate alone was impressive: excluding Internet writings, several hundred essays and five books were published in China and Japan within one year in the subsequent debate.[105]

One reason for the public controversy was the praise showered on Ma and Shi by Japanese media. A prominent Japanese journal quickly translated Ma's essay under the title "Our China: Be Cautious in Opposing Japan" and then published an interview with Ma under the title, "Japan has already apologized to China 21 times."[106] A book under Ma's name appeared in Japan, entitled "Japan has no need to continually apologize to China." Shi Yinhong was invited to visit Japan in April 2003 by the Japanese Foreign Ministry, and his essay was published in a Japanese journal. An interview with both was published in the *Asahi Shinbum* on the anniversary of the end of WWII, August 15, 2003. In response, Ma and Shi were denounced as the *qingripai* (the close-to-Japan faction). Zhuo Nanshen, a Singapore-based journalist, labeled the new thinking as thinly cloaked "old Japanese thinking."[107]

The new thinking debate first spread to popular Chinese media with an extensive article in *Southern Weekend* on June 12, 2003.[108] Widespread newspaper and Internet coverage of scholarly debates quickly engaged the general public.[109] The online debate took off after two essays by *People's Daily* editorial board member Lin Zhibo were posted on the People's Web "China-Japan Forum" in summer 2003. Lin Zhibo ridiculed Ma and Shi personally while describing Japan as a "thief or a hoodlum; a lap-dog cuddling up to its [U.S.] master."[110] Zhou Nanshen criticized Ma and Shi in an edited book, an article in *World Affairs*, and a number of newspaper interviews.[111] Wang Xiaodong, an influential popular nationalist, also gave interviews and issued a critical article in Hong Kong.[112]

Much of the criticism came from an explicitly populist perspective, praising moral righteousness over political pragmatism. Lin Zhibo argued repeatedly that, "Chinese people cannot accept this argument." Wang Xiaodong claimed that since most Chinese people remained unsatisfied with Japan's apologies, the new thinking was untenable.[113] An op-ed in *China Youth Daily* agreed: "As soon as someone tries to raise history issues, they are attacked as 'damaging China-Japan friendship, damaging East Asian peace' and labeled a nationalist. . . . China's average citizens (*laobaixing*) will not accept such methods."[114]

Due to extensive media coverage and active online participation, popular nationalism more quickly dominated the "new thinking" debate than an earlier debate over Japan policy limited to the print edition of *Shijie Zhishi* (World Affairs) or the "China Can Say No" furor of a few years earlier.[115] The Internet also facilitated the public fury released upon Ma Licheng personally. Ma admitted that he received "mountains" of faxes, e-mails, and letters, of which he estimated "at least 70 percent were opposed" to his arguments, calling him a "Chinese traitor" (*hanjian*) and even threatening to kill him.[116] His address and phone number were posted on the web, with calls to burn down his house.[117] In the summer of 2003, Ma requested and took early retirement from the *People's Daily*, left Beijing, and took a job working for Phoenix TV in Hong Kong.[118]

In contrast to the emotional criticism online, a number of leading scholars praised the practical benefits offered by Ma and Shi's approach. Tsinghua University professor Xue Li argued that "Chinese should swallow a bitter fruit, and for the sake of the national interest, seek to overcome the history question."[119] Feng Zhaokui pointed out that "[i]t would have been impossible to improve our relations with Western countries if we had always stressed history."[120] Other scholars offered more qualified support, such as Ling Xingguang's assessment of "Correct strategy, incorrect tactics."[121]

As one review of the debate commented in 2004, "Objectively speaking, given the decline in China-Japan relations, particularly at the popular level, many of the recommendations in the new thinking are quite reasonable. But knowing the disease is not the same as having the right cure." Given the deteriorating state of China-Japan relations, the new thinking was simply too difficult for most people to accept.[122] Jin Xide, one of China's top Japan experts, agreed that the new thinking "presumes that Japanese public opinion is very important, but the Chinese government can simply forcefully suppress China's public opinion." In reality, Jin explained, China's "harsh" stance on history is due to public sentiments in China, which "is not something that one leader or one level of government can simply ignore."[123] Indeed, Foreign Ministry officials closely monitored the Internet debate and met several times with Lin Zhibo and other critics of the new thinking.[124] Yet once the public uproar died down, many of Ma and Shi's proposals did eventually become part of China's Japan policy.

In the wake of the online and street protests in spring 2005 against Japan's entry into the UN Security Council, Chinese leaders and policy experts turned sharply against anti-Japanese popular nationalism. Experts coalesced around a pragmatic criticism of emotional nationalism as destructive of China's overarching national interests. Chinese policy-makers, worried about negative implications on domestic stability and diplomatic relations with Japan, first quickly restrained protests and then began a diplomatic outreach to Japan. By 2007, popular nationalism had been restrained and bilateral relations stabilized. In response, public opinion toward Japan began to improve. The wave of online popular nationalism had come to an end.

Conclusion: Implications and Extensions

The wave of online nationalism toward Japan from 2002 through 2005 undermines traditional assumptions that the spread of the Internet contributes to greater liberal attitudes and policies. Instead, widespread use of the Internet in China for online activism, reading news stories, and public debates strengthened emotional, anti-Japan sentiments, impeded bilateral economic cooperation, and undermined liberal and pragmatic discourses over foreign policy. It was only after the Chinese authorities began to restrict Internet content and online activism that the wave of online nationalism began to subside, after which Chinese foreign policy toward Japan returned to its previous pragmatic trajectory, policy discourse moderated, and public opinion toward Japan began to improve.

One explanation for this surprising outcome may be that a partially free cyberspace, much like a partially free media sector, tends to exacerbate popular nationalism.[125] While intriguing, this counterfactual claim is not supported by the evidence presented above: as the amount of online content and use of the Internet expanded in China, popular nationalism grew stronger. Furthermore, overseas Chinese with unfettered access to the web have played an increasingly central role in China's online nationalism, evident in the online reaction to the Olympic torch protests in spring 2008.[126] It thus seems likely that as the Internet continues to weave its way into the civil society fabric of urban China, the importance of popular, online nationalism will continue to grow.

The Internet also creates a market-like environment in which varying ideas and ideologies must compete for "eyeball hang time" online.[127] Although the Chinese state appears capable of restricting easy access to undesirable content online, it cannot compel Chinese netizens to read, much less agree with, state propaganda. China's pragmatic approach to Japan is forced to compete with liberal and popular nationalist alternatives for the hearts and minds of Chinese netizens. The online playing field, although hardly level, will thus remain a key testing ground among competing ideas shaping public opinion, policy discourse, and foreign policy in China.

Notes

1. Deibert, Ronald J., "Dark Guests and Great Firewalls: The Internet and Chinese Security Policy," *Journal of Social Issues*, Vol. 58, No. 1, January 2002, 143–59; Hachigian, Nina "China's Cyber-Strategy," *Foreign Affairs*, Vol. 80, No. 2, March/April 2001, 118–33; Shanthi Kalathil and Taylor C. Boas, "The Internet and State Control in Authoritarian Regimes: China, Cuba, and the Counterrevolution," *First Monday*, Vol. 6, No. 8 (August 2001), 1, firstmonday.org/issues/issue6_8/kalathil/index.html; Chase, Michael S. and James C. Mulvenon, *You've Got Dissent! Chinese Dissident Use of the Internet and Beijing's Counter-Strategies* (Santa Monica, Calif.: RAND, 2002); Lokman Tsui, "The Panopticon as the Antithesis of a Space of Freedom: Control and Regulation of the Internet in China," *China Information*, Vol. 17, No. 2 (2003),

65–82. For the argument that the internet fosters nationalism in China, see Paul Mooney, "Internet fans flame of Chinese nationalism," yaleglobal.yale.edu/display.article?id=5516, April 4, 2005; Peter Hays Gries, "Tears of Rage: Chinese Nationalist Reactions to the Belgrade Embassy Bombing," *The China Journal* (July 2001): 25–44.

2. Guobin Yang, 2005, "Environmental NGOs and Institutional Dynamics in China." *The China Quarterly* No. 181, 46–66; Guobin Yang, "The Internet and Civil Society in China: A Preliminary Assessment," *Journal of Contemporary China*, Volume 12, Number 36 (August 2003), 453–75; Guobin Yang, 2003, "The Internet and the Rise of a Transnational Chinese Cultural Sphere," *Media, Culture & Society* 25(4): 469–90.

3. The phrase "year of internet nationalism" comes from Liu Ning, "Beijing-Shanghai high-speed rail encounters the threshold of public opinion."

4. Ping Jihua, "Zhongguo dalu shouci minjian baodiao xingdong qiuyuanxin" (A request for support for mainland China's first social action protecting the Diaoyu Islands), at www.1931-9 18.org/bbs/dispbbs.asp?boardID=220&ID=1063 (posted April 8, 2003).

5. For one description of the sign-on campaign's origins and global reach, see Peng Lewu, "1000 Wan qianming de quanliucheng" (The Complete Process of Gaining Ten Million Signatures), *Nanfang Zhoumou* (*Southern Weekend*) (March 31, 2005) [news.sina.com.cn/c/2005-03-31/19286254269.shtml, accessed on 2 November 2009].

6. Suisheng Zhao describes "liberal nationalists" as those who insist on societal autonomy from the state in setting national priorities, such as Wang Xiaodong. He thus includes explicitly illiberal strains, such as nativism, as a thread within liberal nationalism. Suisheng Zhao, "China's Pragmatic Nationalism: Is It Manageable?" *The Washington Quarterly* 29, no. 1 (2005): 133. I prefer to describe the anti-U.S. protests (1999–2001) and anti-Japan protests (2003–2005) as "popular nationalism," reserving the term "liberal" for ideas which adhere to its more classical referents.

7. Interviews in Beijing with Chinese activists (June 2–5, 2008).

8. Feng Jihua, "Zhongguo dalu shouci minjian baodiao xingdong qiuyuanxin" (A request for support for mainland China's first social action protecting the Diaoyu Islands), at www.1931-9 18.org/bbs/dispbbs.asp?boardID=220&ID=1063 (posted April 8, 2003).

9. Li Mutong, "Wangluo minzhuzhuyi xiankai zhongguo minzhuzhuyi xin pianzhang" (Internet Nationalism Opens a New Chapter in China's Nationalism) (September 18, 2003) [news.sina.com.cn/c/2003-09-18/12061767730.shtml, accessed on 3 November 2009].

10. Huang Zhangpu, "The spiral of silence after Yasukuni."

11. They were inspired by a meeting held in Osaka on January 23, 2000: "The Biggest Lie of the 20th Century: the Complete Record of the Nanjing Massacre." See China 918.Net [www.china918.net/918jn/918gkx.htm, accessed on 3 November 2009].

12. See China 918.Net [www.china918.net/918jn/918gkx.htm, accessed on 3 November 2009].

13. He was sentenced to ten months, suspended for three years, after refusing to admit his crime and pay a fee. His work visa was revoked and he was asked to leave Japan within six months. He had been in Japan for nine years by that time. Liu Ning,

"Feng Jihua: Wo yuanzuo yige minzuzhuyizhe" (Feng Jihua: I want to be a nationalist), *Xinwen Zhoukan* (News Weekly), (June 24, 2002): 42–45.

14. Liu Ning, "Feng Jihua." 42.

15. The website is x.1931-9-18.org/. See Fan Shiming, "Burning on the Street, Fuelled on the Net: The Internet and Public Expression in China" (Beijing University, unpublished paper, 2006): 20

16. Han Xuan, et al., "2003: Wangluo minzuzhuyi faren" (2003: Internet Nationalism Explodes and Endures) *Guoji xianfeng daobao* (*International Herald Tribune*) (Beijing) (September 19, 2003): 5. See also the interview of Lu Yunfei in Charles Hutzler, "Yuppies in China Protest via the Web," *Wall Street Journal* (March 19, 2004): 1.

17. Shen Tao, "Shilun dangdai zhongguo wangluo minzuzhuyi jiqi dui zhongri guanxi de yingxiang: yi aiguozhe tongmengwang wei li" (Contemporary Chinese Internet Nationalism and Its Impact on Sino-Japanese Relations: A Case Study of "Patriot Alliance") (master's thesis, Beijing University, June 2006): 36. Studies of social movements find a similar curve of rapid expansion and decline in protest movements. See Tarrow, Sidney, *Power in Movement: Social Movements and Contentious Politics* (Cambridge: Cambridge University Press, 1998): 144–45.

18. Shen Tao, 20–21, 35.

19. As of 2007 at the Alliance of Patriots website, the average age of volunteers was thirty-three, ranging from thirty to forty-five years old. This reflects national trends: 58 percent of all "netizens" in China are male. Shen Tao, 14; 22. A 2002 Chinese netizens survey found that 89 percent of all "netizens" were under thirty-five years old. Center for Humanities and Social Sciences, Chinese Academy of Social Sciences, "Guoqing Diaocha 2002" (2002 Investigation of the National State of Affairs) (Qingdao: Shandong Renmin Chubanshe 2003).

20. Interviews with Patriot's Alliance website volunteers, Beijing, July 12–13, 2007.

21. China Internet Network Information Center (CINIC), "2008 Report on China's Internet Use" [www.cnnic.net.cn/uploadfiles/doc/2008/1/22/212245.doc].

22. A nationwide 2002 survey found that 89 percent of all netizens were below thirty-five years old, with an average age of 26.5 years old. Almost all had a degree beyond high school (89 percent), with 62 percent clustered in three cities: Beijing, Shanghai, and Guangzhou. Center for Humanities and Social Sciences, Chinese Academy of Social Sciences, "Guoqing Diaocha 2002" (2002 Investigation of the National State of Affairs) (Qingdao: Shandong Renmin Chubanshe, 2003): 718, 848.

23. University students' reliance on the Internet for Japan news rose 7.6 percent, television declined 4.9 percent, and newspapers and journals declined 2.4 percent from 2006 to 2007. Li Yu and Fan Shiming, "2007 Zhongri guanxi yulun diaocha baogao" (2007 Report on Public Opinion in China-Japan Relations), Presentation at Third Beijing-Tokyo Forum, Beijing, China (August 28, 2007).

24. "Dui-Ri yulon diaocha," www.comrc.com.cn/zrcb/dryl/2004.htm.

25. "Eyes on the World, Future in Hand: Horizon 2006 Chinese Opinion Polls," Horizon Research Consultancy Group (2006): www.mansfieldfdn.org/polls/poll-06-19.htm. "Societal elites" include academic experts, returnees from overseas, business leaders, media experts, and university students.

26. The middle class indicator is composed of educational, income, and professional attributes. Zhou Xiaohong, "Zhongguo zhongchan jieji diaocha" (Survey of the Chinese Middle Class) (Beijing: Shehui kexue Wensan chubanshe, 2005): 203.

27. The largest number, almost 40 percent, first engage in instant communication ("chatting"). See China Internet Network Information Center (CINIC), "2008 Report on China's Internet Use."

28. The most popular website for international news is Sina.com (picked as most popular among the general public: 45.9 percent; elites: 59.3 percent), followed by Sohu.com (general public: 17.6 percent; elites: 10.5 percent). Xinhua's website was only listed by 4.85 percent of elites and less than 1 percent of the general public; the *People's Daily* website (*renminwang*) was listed by 1.89 percent of elites and less than 1 percent of the general public. "Eyes on the World."

29. It was originally established as the "BBS Forum to Strongly Oppose the NATO Atrocity," and changed its name the following month. See *Qiangguo luntan* [bbs1.people.com.cn/boardList.do?action=postList&boardId=1, accessed on 3 November 2009].

30. The site is the *People's Daily* website (*renminwang*), Japan Channel [japan.people.com.cn/, accessed on 3 November 2009].

31. Interview (Beijing, Institution of Japan Studies, CASS, April 17, 2008). The article is available at the *People's Daily* website (*renminwang*), Japan Channel [japan.people.com.cn/GB/35464/40041/42165/3865467.html, accessed on 3 November 2009].

32. Drawn from Sino Forum [bbs.sina.com.cn/, accessed on 3 November 2009].

33. Drawn from Tianya Forum [www.tianya.cn/, accessed on 3 November 2009].

34. Yan Taining, "Zhongguo guojia liyi yishi jiaozhong de riben: huanqiu shibao ruhe baodao riben" (The National Image of Japan from the view of China's national interest: How does *Global Times* Report on Japan?), *Global Media Journal* (Shijie meiti xuebao) (December 2005): 120–29.

35. "Zhongguo minjian renshi zhi annan huhuxin quanwen piluo" (Revealing the Complete Text of a Chinese Private Individual's Appeal Letter to Anan), *Huanqui* (The Globe), 16 April 2005, 29.

36. The incident is described in Gries, Peter Hays, "China's 'New Thinking' on Japan," *The China Quarterly* 184 (December 2005): 832.

37. The journalist who originally broke the story later explained that "due to the public anger in this case, the Guangdong authorities had to thoroughly investigate this incident." Lin Wei, "Zhu Hai: riben luyoutuan gouchiri daohuan, shengcheng laihua zhiwei maichun" (Zhuhai: Japanese tour group announces that on China's Day of Humiliation, they have come to China just for prostitutes), *China Youth Daily* (September 26, 2003) [news.sina.com.cn/s/2003-09-26/0351820418s.shtml, accessed on 3 November 2009]; *China Youth Daily* was listed as one of the three most widely read papers by Chinese urban elite in a 2006 public opinion poll. "Eyes on the World."

38. Lin Wei, "Ribenke zhuhai maichunlian: shehui zhennu Guangdong lingdao yaoqiu yancha" (Japanese prostitution scandal in Zhuhai: Societal anger pushes Guangdong authorities to seriously investigate), *China Youth Daily*, 28 September 2003 [news.sina.com.cn/s/2003-09-28/03471829615.shtml, accessed on 3 November 2009].

39. For instance, see chatroom comments at Di Er Channel BBS [www2.nedu.edu.cn/bbs/archiver/?tid-2461.html, accessed on 3 November 2009].

40. Zhao Guochen, "Cong zhongri minjian shijian tuoshi wangluo minzuzhuyi" (Perspectives on Chinese Online Nationalism Through a Series of Civil Society Incidents between China and Japan) (MA thesis, China Media University, 2005): 18–19.

41. "800 Wan wangyou qianming fandui riben cangai jiaokeshu" (8 Million Netizens Sign Up to Protest Japan's Revised Textbooks), *Jinghua Shibao* (April 7, 2005): 1.

42. Zhao Guochen, "Perspectives on Chinese Online Nationalism," 20.

43. Interview (Beijing, Institution of Japan Studies, CASS, April 17, 2008).

44. The percentage is slightly higher among the middle class who are "within the government system" (such as officials and staff of state-owned companies) than those "outside the system." Zhou Xiaohong, "Zhongguo zhongchan jieji diaocha" (Survey of the Chinese Middle Class) (Beijing: Shehui kexue Wensan chubanshe, 2005): 322.

45. Center for Humanities and Social Sciences, Chinese Academy of Social Sciences, "Guoqing Diaocha 2002" (2002 Investigation of the National State of Affairs). (Qingdao: Shandong Renmin Chubanshe, 2003): 919.

46. Chen Weixing, *Chuanbo de guannian* (Perspectives on Broadcasting) (Beijing: Renmin Chubanshe, 2004): 252.

47. Li Ying, "Zhongguo xuyan shenmeyang de minzuzhuyi" (What kind of nationalism does China need), *Guoji xienfeng daobao* (International Herald Tribune) (Beijing) August 4, 2003, at Sino.com [news.sina.com.cn/c/2003-08-04/15541475532.html, accessed on 3 November 2009]. For a critical response, see Lin Jing, "Wangluo minzu zhuyi/Yihetuan xixiang" (Internet Nationalist is not Equivalent with the Taiping Rebels' Thinking), *Guoji xienfeng daobao* (International Herald Tribune) (Beijing) (September 19, 2003): 1.

48. See, respectively, Le Shan, ed, *Qianliu: dui xaiai minzu zhuyi de pipan yu fansi* (Undercurrent: Criticism and Reflections on Parochial Nationalism) (Shanghai: Huadong shifan daxue chubanshe, 2004); Wang Guoping, "Cong fandui ribe changren kan minjian xingwei de zuoyong" (Understanding the Role of Social Action from the Opposition to Japan's Permanent Membership in UN Security Council), *Huanqiu* (The Globe), April 16, 2005, 28–29.

49. He was interviewed by South Korean media, later translated in Yi Zhengli, "Xuezhe xinlun: Fangzhi jiduan minzuzhuyi qingxu 'jiechi' zhongguo" (New Scholarly Forum: Oppose Extreme Nationalist Emotions Which "Hijack" China), *Renminwang* (People's Daily Online) (January 8, 2004) [news.sina.com.cn/c/2004-01-08/08482575974.shtml, accessed on 3 November 2009].

50. Lifeng Jiang, "Zhongguo minzhong dui riben henshao you qingjingan: diyici zhongri yulun diaocha jieguo fenxi" (Chinese public rarely feels close to Japan: analysis of the results of the first China-Japan public opinion survey), *Riben Xuekan* (Japan Studies), 6 (2002): 1–18; Lifeng Jiang, "Zhongguo minzhong dui riben de buqingjin xianzhu zengqiang" (Chinese public's feeling of not close to Japan noticeably strengthens), *Riben Xuekan* (Japan Studies), 6 (2004): 6–32; Lifeng Jiang, "Peiyu liangguo renmin de qingjingao dui gonggu zhongri youhao de genji yiyi zhongda" (Cultivating feelings of closeness between people of the two countries has great sig-

nificance for consolidating the foundation of friendship between China and Japan), *Riben Xuekan* (Japan Studies), 6 (2006): 1–28.

51. Li Yu and Fan Shiming, "2005 zhongguo wuge chengshi gongmin duiri yulun diaocha baogao" (2005 Report on Chinese Public Opinion toward Japan in Five Chinese Cities), unpublished report (2005): 7.

52. The first two are from the 2002 and 2004 CASS surveys (see note 79); the last one is from Japan Research Center, *Nicchu Kankei ni tsuite no Kokusai Hikaku Yoron Chosa Kekka* (Comparative Survey Results in Sino-Japan Relations), July 2005; cited in Mindy L. Kotler, Naotaka Suguwara, and Tetsuya Yamada, "Chinese and Japanese Public Opinion: Searching for Moral Security," *Asian Perspective* 31, no. 1 (2007): 93–126.

53. Li Yu and Fan Shiming, "2005 Report on Chinese Public Opinion."

54. The rate of response for those that agreed that visits are acceptable under the condition that Japan apologizes satisfactorily for the wartime invasion or removes the souls of Admiral Tojo and other war criminals are as follows: 2002—31.6 percent; 2004—36.3 percent; 2006—41.7 percent. The response rates for those who accepted no visits under any conditions: 2002—50.9 percent; 2004—42.0 percent; 2006—45.3 percent. See Lifeng Jiang, 2002, 2004, 2006.

55. Lifeng Jiang, 2002, 2004, 2006.

56. Lifeng Jiang, 2002, 2004, 2006.

57. Snyder, Jack, and Karen Ballentine, "Nationalism and the Marketplace of Ideas," *International Security* 21, no. 2 (Autumn 1996): 32.

58. Drawn from interviews 24 (May 2007), 38 (July 2007), and 59 (September 2007).

59. Johnston, Alastair Iain, "Chinese Middle Class Attitudes Toward International Affairs: Nascent Liberalization?" *The China Quarterly* 179 (September 2004): 626.

60. Cited in Zhang Chongfang and Wang Jianhua, "Zhong guo waizhang shouci tongguo hulianwang yu wangmin jiaoliu waijiao wenti" (Chinese Foreign Minister's first interaction with netizens on foreign affairs via the Internet), December 23, 2003, published online at news.rednet.com.cn/Articles/2003/12/503488.HTM. "Netizen" (*wangmin*) refers to Internet users.

61. Market analysts expected that China would prefer to use a single type of technology for all its high-speed trains. The first round of the project included building train cars and constructing the Beijing-Shanghai and Shanghai-Hangzhou lines. See Liu Ning, "Jinghu gaotie zaoyu minyi menjian" (Beijing-Shanghai High-Speed Rail Encounters the Threshold of Public Opinion), *Nanfeng chuang* (Southern Wind Window), August 16, 2003 [news.sina.com.cn/c/2003-08-16/14521556454.html, accessed on 3 November 2009].

62. Liu Ning, "Beijing-Shanghai High-Speed Rail."

63. Liu Ning, "Beijing-Shanghai High-Speed Rail."

64. Ming Wan, *Sino-Japanese Relations: Interaction, Logic, and Transformation.* (Stanford, Calif.: Stanford University Press, 2006): 369n63.

65. Liu Ning, "Beijing-Shanghai High-Speed Rail."

66. Interview with a Nanjing University professor (Hong Kong, August 24, 2006).

67. Han Xuan, "Hujing gaotie pengshang minzuzhuyi liangchao" (The Beijing-Shanghai High Speed Train Encounters a Wave of Nationalism), *Guoji Xianfeng Daobao* (International Herald Tribune [Beijing]) August 4, 2003 [news.sina.com.cn/c/2003-08-04/15541475533.html, accessed on 3 November 2009].

68. Przystup, "Japan-China Relations: Not Quite All about Sovereignty—But Close."

69. "Riben zhengduo jinghu gaotie qianjing buming; faguo lungui jiantao kai-jiang" (Prospects Unclear for Japan's Pursuit of the Beijing-Shanghai Rail Line; France's Rail Technology Begins to Take the Lead), *Financial Times* (Caijing shibao), January 31, 2004 [news.sina.com.cn/c/2004-01-31/07582730309.shtml, accessed on 3 November 2009].

70. Han Xuan, "The Beijing-Shanghai High Speed Train."

71. Han Xuan et al., "2003: Wangluo minzuzhuyi faren" (2003: Internet Nationalism Explodes and Endures), *Guoji xianfeng daobao* (*International Herald*), September 19, 2003, 5. Japanese companies later speculated that perhaps the German or French competitors encouraged the spread of this news to the Chinese public. See Liu Ning, "Beijing-Shanghai High-Speed Rail," 44.

72. The sign-on campaign (no longer available) was at www.1931-9-18.org/bbs/guest_sign.ASP.

73. Han Xuan, "The Beijing-Shanghai High Speed Train."

74. A search using the Google search engine in China at the time revealed 20,200 hits when searching for the sign-on campaign. Shen Tao, "Contemporary Chinese Internet Nationalism," 24.

75. Gries, "China's 'New Thinking,'" 844.

76. See, for instance, Niu Jianjun, "Riben jiaotongchen wugong erfan tulao daikuan huan jinghu gaosu keneng tulao" (Japan's transportation slavery-getting benefits without merit: using loans to get the Beijing-Shanghai high-speed rail may be labor in vain), *Financial Times* (*Caijing shibao*), August 11, 2003; and coverage in *Cankao Xiaoxi* (Reference News) (December 31, 2003), 8.

77. Shi Hongtao, "Ribenren dui zhongguo jinghu tielu gan xingqu, jiji tuixiao xinganxian" (Japanese people have interest in China's Beijing-Shanghai rail line and actively promote the *shinkansen*), *Beijing Qingnian Bao* (Beijing Youth News), August 11, 2003 [news.sina.com.cn/w/2003-08-11/11181520103.shtml, accessed on 3 November 2009].

78. Liu Ning, "Beijing-Shanghai High-Speed Rail," 44.

79. Li Mutong, "Wangluo minzhuzhuyi xiankai zhongguo minzhuzhuyi xin pian-zhang" (Internet Nationalism Opens a New Chapter in China's Nationalism), *Guoji xianfengbao* (International Herald Tribune-Beijing), September 18, 2003 [news.sina.com.cn/c/2003-09-18/12061767730.shtml, accessed on 3 November 2009].

80. Yang Jinlin, "Zhongguo zhengzai xingqi xinyilun minzuzhuyi liangchao?" (Is a New Wave of Nationalism Now Springing Up in China?), *Nanfeng Chaung* (Southern Window), October 16, 2003 [news.sina.com.cn/c/2003-10-16/19161934982.shtml, accessed on 3 November 2009].

81. Li Mutong, "Wangluo minzhuzhuyi."

82. "Alstom, Kawasaki Heavy, Bombardier win China high-speed train contracts," *Xinhua Financial Network News* (August 30, 2004). The total Japanese order was valued at approximately 140 billion yen (US$1.3 billion), with Kawasaki Heavy Industries' portion projected to be around eighty billion yen. Kawasaki would initially manufacture the trains in Japan, with China's CSR Sifang Locomotive and Rolling Stock Co., the Chinese partners in the consortium, taking over production at a later date. See "Kawasaki Heavy Clinches Chinese Rail Car Order," *Asia Pulse* (October 21, 2004).

83. Shen Tao, "Contemporary Chinese Internet Nationalism," 33.

84. "Chinese Officials Shut Down Anti-Japanese Activists' Web Site," Kyodo (August 31, 2004).

85. See Jane Cai, "Patriot's Website Closed Because of Railway Protest," *South China Morning Post* (September 1, 2004): 5; "China shuts down anti-Japanese website for opposing rail deal," Agence France Presse (September 1, 2004); "Chinese Officials Shut Down Anti-Japanese Activists' Website," BBC Monitoring International Reports (August 31, 2004). "Aiguozhe tongmengwang zhige xiongde wangzhan de ganxie xin" (Patriots Alliance web issues a thank-you letter to all brother websites) (September 4, 2004, at www.sinophoto.com.cn/content/040912/04091226.htm).

86. Niu Jianjun, "Rijun duji shijian suochang jaijin jiaoshe; minjian huodong shengsuan jiaoda" (Bargaining Tightens in the Demands for Compensation from the Incident of Japan's Military Poisoning; A Relative Success for Civil Activism), *Financial Times* (Caijing shibao) (September 6, 2003) [news.sina.com.cn/c/2003-09-06/07391686999.shtml, accessed on 3 November 2009].

87. Contracts on the high-speed rail project were awarded to French companies in October 2004, and to German companies on June 7, 2005, and on June 6, 2005. See Zhang Xufeng, "Shei xian shixiang jinghu gaotie" (Who Will Drive the Beijing-Shanghai High Speed Train First?), *Hubei Daily* (*Hubei Ribao*), April 21, 2006 [news.sina.com.cn/c/edu/2006-04-21/10088755344s.shtml, accessed on 3 November 2009]; "China Has Not Initiated Beijing-Shanghai Express, MOR," *International Financial News* (June 9, 2005): 1.

88. *Taipei Times* (December 31, 2004).

89. "China denies plan to order Japanese bullet trains," Kyodo News Service (November 23, 2005).

90. The decision to use mag-lev technology was widely seen as a victory for the Shanghai city government, which owns the Shanghai Mag-Lev Transport Co. (SMTC) that will get most of the work, although Transrapid, a German consortium, is likely to get substantial contracts as well. Mark O'Neill, "Decision on high-speed rail throws foreigners off track," *South China Morning Post* (March 20, 2006): 6.

91. In fact, the ministry required only that 70 percent of the rail project content must be from Chinese companies. "Zhongguo gaotie mengxiang sange tielu jishu zuifada guojia qiangduo shichang" (China's High-Speed Rail Dream: Three of the Most Developed Countries With High Technology Plunder the Market), *Zhongguo keji caifu* (China Science, Technology, and Fortune), July 3, 2006, finance.sina.com.cn/g/20060703/17342701174.shtml; "Jinghu gaotie you wangniannei kaigong" (Work

on the Beijing-Shanghai High-Speed Train Will Commence Within the Next Year), *Yangzi Evening News* (October 25, 2006).

92. From September 2002, when Japanese firms seemed likely to receive contracts, through the final decision that Chinese firms would build the rail line, announced in March 2006, Prime Minister Koizumi visited the Yasukuni Shrine four times: on April 21, 2002; January 14, 2003; January 1, 2004; and October 17, 2005.

93. Ming Wan, 256.

94. During a televised news conference on March 14, 2006, at the close of the National People's Congress, Premier Wen Jiabao pledged China's willingness to meet with any Japanese Prime Minister who did not visit the Yasukuni Shrine. Premier Wen also called for an expansion of economic ties and people-to-people exchanges to build mutual trust and cooperation for "win-win results." *Renmin Ribao* (April 1, 2006), *The People's Daily* [politics.people.com.cn/BIG5/1024/4260505.html, accessed on 3 November 2009].

95. Interviews, Railway Ministry officials, Beijing, March 22, 2006.

96. For instance, Jin Xide, a prominent Japan expert at CASS, had publicly supported Japanese involvement in the project to foster trust and cooperation. See Han Xuan et al., "2003: Internet Nationalism Explodes," 5.

97. Ma Licheng and Ling Zhijun, *Jiaofeng* (*Crossed Swords*) (Beijing: Jinri Zhongguo chubanshe, 1998); Ma Licheng and Ling Zhijun, *Huhan* (*Outcry*) (Guangzhou: Guangzhou Chubanshe, 1999). See also Lu Yi, *How Far Away Is Mutual Understanding*, 328; Joseph Fewsmith, "Crossed Swords: A True Account of The Three Periods of Ideological Liberation," *Foreign Policy*, 113 (Winter 1998/1999): 107–10.

98. Ma Licheng, "Duiri guanxi xinsiwei" (New Thinking on relations with Japan), *Zhanlue yu guanli* (Strategy and Management), 6 (2002): 41–47.

99. Shi Yinhong, "Zhongri jiejing yu 'waijiao geming'" (China-Japan rapprochement and a diplomatic revolution), *Zhanlue yu guanli* (Strategy and Management), 2 (2003): 71–75.

100. Shi Yinhong, "Lishi wenti yu dazhanlue quanheng" (The history question and the great strategic balance), *Kangri zhanzheng yanjiu* (Journal of the War of Resistance to Japan), No. 3 (2003): 191–95.

101. Zhaokui Feng, "Lun duiri guanxi xinsiwei" (Discussion of the new thinking on relations with Japan), *Zhanlue yu guanli* (Strategy and Management), 4 (2003): 4.

102. After a fierce closed-door debate over Japan policy, the final Party Congress Resolution declared that China would "Treat neighbors as friends; treat neighbors as partners" (*yilin weishan; yilin weiban*). China's Japan experts widely agreed that this policy signaled the new leadership's intent to strengthen ties with Japan, an analysis reinforced by Zeng Qinghong's encouraging comments to the chairman of the Japan-China Friendship Association on November 22. Interviews at Beijing University (May 15, 2007); Renmin University (August 12, 2007). See also Shi Jiafang, *Zhanhou ZhongRi guanxi: 1945–2003* (*Postwar China-Japan Relations: 1945–2003*) (Beijing: Dangdai chubanshe, 2005): 219; Isao Yamamoto, "Japan urgently needs to rethink nature of relationship with China," *The Nikkei Weekly* (Japan) (November 17, 2003).

103. Hiroyuki Sugiyama, "*People's Daily* slams anti-Japan rhetoric," *Yomuri Shinbum* (December 14, 2002).

104. Interviews, Beijing (April 6, 2007; March 14, 2007; September 30, 2007).

105. Liu Ning, "One year anniversary," 18.

106. The journal is *Bungei shunju*, issues 3 and 10; cited in Lu Yi, *Zhongri xianghu lijie haiyou douyuan? Guanyu liangguo minzhong xianghu renshi de bijiao yanjiu* (How Far Away Is Mutual Understanding Between China and Japan? A Comparative Study on the Mutual Perceptions Between the Two Peoples). Beijing: World Knowledge Press, 2006, 318.

107. Nanshen Zhuo, "Jiujing shui xuyao xinsiwei" (Who exactly needs the new thinking), *Shijie zhishi* (World Affairs), 15 (2003): 44–46.

108. Sun Yafei, "Duiri xubuxuyao 'xinsiwei'?" (Is the "new thinking" toward Japan really necessary), *Nanfang Zhoumou* (*Southern Weekend*), June 12, 2003.

109. Special editions on the "new thinking" were issued by *Shijie zhishi* (World Affairs), *Kangri zhanzheng yanjiu* (Journal of the War of Resistance to Japan), and *Zhanlue yu guanli* (Strategy and Management). Prominent conferences on the new thinking held in Beijing in July and August 2003 were widely covered in the Chinese press. See Wang Yizhou, "Zhongri guanxi de shige wenti" (Ten issues in China-Japan relations), *Shijie jingji yu zhengzhi* (*World Economics and Politics*) (September 2003): 8–9.

110. Lin Zhibo, "Duiri guanxi xinsiwei zhiyi" (Questioning the "new thinking in China-Japan relations") (July 22, 2003), *The People's Daily* [www.people.com.cn/BIG5/guandian/1036/1978531.html, accessed on 3 November 2009]; Lin Zhibo, "Duiri guanxi xinsiwei zai zhiyi" (Another question on the "new thinking in China-Japan relations") (August 15, 2003), *The People's Daily* [www.people.com.cn/BIG5/guandian/183/8456/8457/2017423.html, accessed on 3 November 2009]; Lin Zhibo, "Dangdai zhongguo shifou xuyao minzuzhuyi" (Does contemporary China need nationalism) (December 21, 2004), *The People's Daily* [opinion.people.com.cn/BIG5/1036/3070502.html, accessed on 3 November 2009].

111. Nanshen Zhuo, "Who exactly needs the new thinking."

112. Xiaodong Wang, "21 shiji zhongri guanxi—yichang youjiang jingcai youxi" (Sino-Japanese relations in the 21st Century—A Lottery Game?), *Hong Kong Commercial Daily* (February 24–25, 2003): A07.

113. Xiaodong Wang, "Sino-Japanese relations in the 21st Century."

114. Shi Hongtao, "Beijing bufen xuezhe zhiyi duiri xinsiwei" (Some Beijing scholars raise questions about the new thinking toward Japan), *Zhongguo qingnianbao* (China Youth Daily), September 29, 2003.

115. For this debate, see Li Zhaozhong, "Zai Lishi wentishang: ribenren ruhe kandai zhongguo" (How do Japanese people look at China on the history question), *Shijie Zhishi* (World Affairs), No. 1 (January 2000): 33; Wei Zhong, "Butubukuai" (Uncomfortable if keeping silent), *Shijie Zhishi* (World Affairs), 7 (2000): 25; Li Zhaozhong, "Huida yizhong jiandian de wujie" (To answer a kind of simple misunderstanding), *Shijie Zhishi* (World Affairs), 8 (2000): 28–29; and letters to the editor in *Shijie Zhishi* (World Affairs), 2000, editions 1, 12, and 13.

116. Sun Yafei, "Is the 'new thinking' toward Japan really necessary."

117. Liu Xiaobo, "Hezhe aiguo, hezhe haiguo?" (Who loves their country and who harms their country?), *Waitan huabao* (The Bund), August 14, 2003.

118. Ma returned to China in 2006 and began writing a weekly column in *Jingji Guancha* (Economic Observer) the leading liberal economics newspaper in China, indicating that the furor had finally blown over.

119. Xue Li, "Zhongri guanxi nengfou chaoyue lishi wenti?" (Can Sino-Japanese relations overcome the history question?), *Zhanlue yu Guanli* (Strategy and Management), No. 4 (2003): 28–33.

120. Feng Zhaokui, "Lun duiri guanxi xinsiwei" (Discussion of the new thinking on relations with Japan), *Zhanlue yu guanli* (Strategy and Management), 4 (2003): 1–17. See also: Feng Zhaokui, "Duiri guanxi de jiannan quisuo" (A difficult search in China's relationship with Japan), *Shijie jingji yu zhengzhi* (World Economics and Politics), 5 (2004): 26–31; Feng Zhaokui, "Guanyu ZhongRi guanxi de zhanlue sikao" (Strategic thoughts on Sino-Japanese relations), *Shijie jingji yu zhengzhi* (World Economics and Politics), 11 (2000): 12.

121. Xinguang Ling, "Zhanlue duitou zhanshi quiantuo—ping Ma Licheng he Shi Yinhong de liangpian wenzhang" (Correct strategy, incorrect tactics—Comments on the two articles by Ma Licheng and Shi Yinhong), *Shijie jingji yu zhengzhi* (World Economics and Politics), 9 (2003): 17–21; see also Lu Zhongwei, "Zhongri guanxi: lijie yu tuijing" (Sino-Japanese relations: understanding and enhancing), *Xiandai guoji guanxi* (Contemporary International Relations), 9 (2003): 1–8.

122. Liu Ning, "One year anniversary," 18.

123. Quoted in Sun Yafei, "Is the 'new thinking' toward Japan really necessary."

124. Interviews, Institute of Japan Studies, CASS, September 28, 2007; Renmin University of China, September 30, 2007.

125. For the impact of a partially free media sector, see Jack Snyder, *From Voting to Violence, Democratization and Nationalist Conflict* (New York: Norton, 2000): 60.

126. See, for instance, Matthew Forney, "China's Loyal Youth," *The New York Times* (April 13, 2008), *New York Times* [www.nytimes.com/2008/04/13/opinion/13forney.html?em&ex=1208318400&en=45f3cfe648f924cb&ei=5087%0A, accessed on 3 November 2009]; Ariana Eunjung Cha and Jill Drew, "New Freedom, and Peril, in Online Criticism of China," *The Washington Post* (April 17, 2008), at [www.washingtonpost.com/wp-dyn/content/article/2008/04/16/AR2008041603579_pf.html, accessed on 3 November 2009].

127. The (American) football analogy comes from William Safire, "On Language: The Way We Live Now," *The New York Times Magazine* (August 8, 1999).

4

Networking Anti-Japanese Protests

*Popular Sovereignty Reasserted since 2005**

Shih-Diing Liu

IN APRIL 2005, there appeared a series of anti-Japan mass rallies in South Korea and China almost simultaneously. The demonstrators in both countries were mainly protesting against Japan's recent justification and glorification of the Japanese empire's colonial rule in Asia, as well as the Japanese government's bid for permanent membership on the UN Security Council. In China, what is particularly dramatic about the mass protests may be the significantly growing use of the Internet and personal mobile phones for spreading protest information and mobilizing street direct action among the Chinese public. The protests—which were fueled by the immediate proliferation of protest information and the sporadic calls for action through websites, bulletin boards, e-mail, instant online messaging, and text-messaging on mobile phones—have been seen as the "possible harbinger of future popular movements" in China.[1]

In recent years, the Internet has been dramatically used by the Chinese people for mobilizing various kinds of protest action or political campaigns.[2] The information and communication technology provides a new means for the organization of Chinese popular protests from below, and reconfigures the way in which Chinese popular resistance takes shape. Despite some official restrictions and prohibitions,[3] the Chinese protesters were still able to make their way to the Internet and disseminated a range of protest information extensively across the whole nation. Compared with the Chinese state-run media, the booming online public forums, bulletin boards, and chatrooms offer up an alternative political space for the Chinese people to express, discuss, and share their views about public affairs, which was not possible before the appearance of the new media in China.[4]

In this chapter, I will try to illustrate the dynamics of recent anti-Japan movements, by revealing the bottom-up, autonomous "network struggles" (*wanglu kangzheng*) on the Internet. I shall pay special attention to the appearance of two instances of the network struggle during March and April 2005: one is the struggle over meaning with regard to the state repression over the protest; the other revolves around the boycott campaign against Japanese goods. These struggles should be seen as a continuation from the preceding anti-Japan rallies on the Internet.[5] The study shall concentrate its focus on the discursive contestation and negotiation on the Strong Nation Forum (*qiangguo luntan*, at www.qglt.com/bbs/start), a Chinese-language online forum affiliated with the state press *People's Daily*, which I have observed as a participant since the beginning of 2005. The forum was formally created during the 1999 bombing of the Chinese embassy by a U.S. warplane in Belgrade, when the chatrooms provided the Chinese people with an outlet for delivering their anger and engaging public debate. Since then, the forum has become one of the most dramatic and controversial virtual political forums in China. I will concentrate my discussion on the period of March and April 2005, when the protests against Japan's UN bid broke out and the issue was most fiercely discussed and debated on the Internet. The main object of my analysis is the China-Japan forum (at www.qglt.com/bbs/chbrd?to=18), one of the most popular forums managed under the Strong Nation Forum. The China-Japan forum is a public platform especially created for the discussion of a variety of China- and Japan-related affairs, and during the outbreak of the nationalist protests, lots of interesting debates were taking place on it.

The Discourse of Dissent: Contesting Official Narratives

In cyberspace, the Chinese nationalists are keen to express and defend their own views about the nation. Despite some postings considered to be too extreme or politically sensitive and thus deleted by "forum masters" (*banzhu*), we can still find some entries that implicitly or explicitly question or challenge the official narrative disseminated by the state or municipal media such as *People's Daily*, *Xinhua News Agency*, and *Liberation Daily*. Throughout the whole month of April 2005, when the protest against Japan's bid for the UN Security Council membership began to evolve and break out on a nationwide scale, the Internet was used by the Chinese nationalists to create their own meanings and interpretation frames over the protest. The information, messages, and opinions posted on the forum were often in conflict and contradiction with that of the Chinese official authority. Let me raise some examples.

On 7 April, there appeared a posting questioning why the China Central Television (CCTV) and other state media did not cover the protest taking place in Shenzhen.[6] In an entry entitled "On the protest against Japan in Shenzhen" appearing the same day, a participant criticized the absence of news coverage on the protest:

> The people in Shenzhen have come to the forefront of resistance against Japan. In capital Beijing, China's political, economic and cultural center, the protest is yet to begin. What is more unfortunate is that the relevant articles and videotaped report circulated on the Internet are for well-known reason not available to a wider range of citizens. . . . Moreover, it is said that the Central Propaganda Department has assigned the control over the news coverage on the protest. As righteous citizens, how should we understand the doing of the Central Propaganda Department, and the theory which claims that the Chinese Communist Party represents the most fundamental interest of the majority of people?

The government control over the news coverage was confirmed by another participant, who claimed that the source of information came from Japanese news media. In addition to the denouncement of the Chinese state-run media, some forum participants also tried to analyze the news items announced by the websites affiliated with the state media and compare them with other reliable news sources. In an entry dated 12 April, a participant posted a brief but sharp analysis on how the state media Xinhua News Agency attempted to orchestrate the message supporting Japan's bid for the UN Security Council membership. The analysis was then followed by another group of postings that charged the Xinhua News Agency with "misguiding the Chinese public." Another posting then compares the same news topic made by Xinhua News Agency and Singapore's *United Morning Daily*. A participant, who posted a piece of news coverage made by the Japanese media Asahi.com, questioned "why the Chinese media still pretend to be deaf and dumb." Others then suggested the forum participants obtain "genuinely true news and thoughts" from another commercial website in China. On 18 April, an entry appeared on the forum objecting to the perception of the peaceful protest as violence and riot. On 22 April, by referring to the Japanese news media's coverage of a Japanese congressman's visit to the Yasukuni Shrine, a participant complained that the Chinese media did not cover the same event. In addition, there was also an analysis of a local TV station's pro-Japan stance.

There is always the latest news and up-to-date information regarding China-Japan relations on the forum. On 17 April, when the *People's Daily* featured an article entitled "Sustaining stability through the construction of a harmonious society" on its front page with a message to cool down the protests, the forum participants were apparently annoyed by the article's view and expressed their collective resentment toward the official message. The criticism over the article lasted for three days and many participants seemed very unhappy about the attitude of the Chinese government:

> Should a "harmonious" society that is being assaulted and bullied still be called "harmonious"? We'd rather not have such a "harmonious" society.
>
> Concentrating on the development is just a utopian self-fantasy. The real situation would not evolve according to your own imagination. Foreign antagonistic forces would never give up their ambition . . . they would never allow you to develop yourself peacefully and stably. This is just a self-deceiving slogan fooling ordinary people!

> Be watchful of the collusion that takes the construction of harmonious society as an excuse but surrenders and betrays the national interest!

Despite there being fierce criticism over the Chinese government's moderate attitude toward Japan, on the consecutive two days (18 and 19 April), some forum participants who supported the government posted an editorial of *People's Daily*, entitled "Holding a wide breadth of vision, maintaining stability," on the forum. The main idea of the editorial can be summarized as follows: Chinese people should treasure and maintain the great achievement of the country's economic development, which is the "fundamental interest of the Chinese nation." In criticizing the escalation of Japanese militarism and the Japanese right-wingers' plot to distort the history of Japanese colonial imperialism, the editorial demands Chinese citizens "uncompromisingly support the stance of our country's government and policy toward Japan, and believe in the party and the government's capability in dealing with the complex situation and resolving the complex international problem adequately."

The posting of the editorial on the forum, however, quickly stirred up a flame war that, when more official announcements like this were put up against the popular protest, lasted for a couple of days until the government decided to formally crack down on the demonstrators. The dispute was quickly divided into two major camps: for or against the government. For those who supported the government's moderate stance, there was a commonly shared sense of fear over the likely social unrest and disorder as a consequence of the nationwide mass protests. As for those who opposed the government's standpoint in this instance, the notion of "stability," "harmony," and "development" is just another pretext or excuse used to sustain the interest of the status quo:

> This is a typical historical nihilism! Actually, among those who ignore history, Chinese people should be blamed most! Chinese people should take the history

of humiliation and the fundamental cause of such a history seriously. Don't make the same mistake any more!

The [government's] cliché has been sung for tens of years.

Because the government has not openly opposed Japan, we don't need to oppose Japan. It doesn't make sense.

Something wrong with your brain! Just look at South Korea!

I'm a Chinese. I wish I can see a true anti-Japan [stance].

Only a posting or a text message can call upon tens of thousands of people while defying the law.

How bodacious you are!

Don't pretend to be worried about the country. The old China was destroyed by the insensitive people like you.

On the following couple of days, the forum was fueled with the debate over the government's stance. In the fierce contestation of different views, the pro-government postings seemed to have taken the initiative in shaping up its campaign against those participants who advocated a more confrontational stance against Japan. Perhaps because the political atmosphere became more sensitive and uncertain than in earlier periods, the voice of the oppositional camp seemed to get weaker. Even so, at this moment, the protests had already taken place on the streets of Shanghai, Beijing, Shenzhen, Guangzhou, Chengdu, and other major cities across China. Against the rising tide of popular movements, the government had also taken some measures trying to suppress the demonstrations. The "425 incident," in this sense, then marks the split-up of state-led and popular nationalisms.

On 25 April, however, *Liberation Daily*, the Shanghai municipal Communist Party Committee's official newspaper, published an editorial statement on its front page, which used an unusually severe phrase denouncing the recent protests and the protesters. Since the outbreak of the anti-Japan rallies in early April, this is the first time the Chinese authority issued a formal warning against its people. The announcement, entitled "Understanding the essence, investigating the violation of law," explicitly accused the demonstrations of being illegal and therefore must be halted immediately. "A vast body of facts," the editorial wrote, "has shown that the demonstrations that occurred recently are not patriotic activities at all, but illegal conduct; they are not activities participated in by the citizens voluntarily, but are backed up by certain conspiracy." Immediately after the publication of the official warning, lots of criticism and anger appeared on the China-Japan forum. Some participants began complaining about the article and the paper, posting up their views that were full of anger and confusion:

> Don't label the protest as "irrational"—the students and the mass will object it!
>
> The performance of the news media is terribly bad!
>
> This article is the shame of Shanghai!
>
> Will the teenagers accused by the article still have patriotic passion in the future? Why couldn't [the government] find a better solution?

The feeling of discontent and resentment toward the state media and the government suddenly erupted, and the expression of anger proliferated rapidly with more and more participants becoming aware of the news. One participant then explained why she/he felt so unhappy and frustrated about the *Liberation Daily*'s news comment:

> Some people, including some government leaders, simply call the protests as anti-Japan demonstrations, while I think such a notion is inappropriate. Generally speaking, the nature of the protest should be understood as a voluntary mass action against the Japanese right-wingers' falsification of history, a patriotic action against the right-wingers' interference in China's domestic affairs. This is not indiscriminately anti-Japan. Next, overall speaking, the entire protest was carried through with rationality and calmness and no large-scale disorder occurred. . . . Based on the above two points, I think that Shanghai's media has made a serious mistake over the reportage of the protest: First, it did not report the protest from the beginning . . . as a result the whole news report on the protest was dominated by foreign media. Second, the news report was seriously biased. It did not pay attention to any positive side of the protest; instead of report on the patriotic passion of the majority of the participants, the media deliberately exaggerated the vandalism committed by a few mobs. The report just echoed with the foreign media and misguided the international society. It has produced a very negative influence. Third, [the media and the government] did not encourage the young students' enthusiastic patriotism and lead them to a more rational way, but rather accused them of damaging the stability of overall situation and committing street politics. This was just frustrating everybody's enthusiasm. . . . Therefore, if I were to give a mark for the performance of Shanghai's media, I think it would have definitely failed.

Other participants then compared the *Liberation Daily*'s editorial practice with that of the *People's Daily* during the 1989 Tiananmen popular social movement:

> Show your proof on the alleged "backstage manipulation"! It's just the same as 1989!

> I was very shocked after reading the combative article . . . how much negative impact will this article have on the mentality of those involved in the anti-Japan demonstrations? One of the main factors of the escalation of the June Fourth student strikes [in 1989] was the labeling of the protests as antiparty and anti-socialism by the editorial [of *People's Daily*] on 26 April. Has the Shanghai municipal government ever tried to figure out the consequence of the report?

The forum participants continued the attack on the official statement as well as other pro-state narratives in the aftermath of the 425 incident. Despite the popular protest on the streets being suppressed by the state, the strength of the demonstration on the Internet seemed to become much more vigorous and energetic than before. The forum's earlier preoccupation with Japan's war crimes was then replaced by a series of raging criticisms against the Chinese government (including the municipal and central levels) and the state media over the official measure taken to prohibit the anti-Japan demonstration from growing larger. On 29 April, some postings came out to ask the *Liberation Daily* to apologize to the Chinese people; the China-Japan forum soon became a battleground where the stance of the Chinese government and the state-run media was fiercely criticized by the forum participants.

The bottom-up, spontaneous cyberprotest against the official discourse shows that the 2005 anti-Japan rows cannot be simply dismissed as top-down government manipulation or party propaganda, as some Western "China experts" have proclaimed.[7] Rather, the Chinese cybernationalists are keen to find their way to engage in nationalist politics, and claim for the nation a vision that is not necessarily in line with the official discourse. The Internet, with its decentralizing and participatory nature, helps facilitate a bottom-up force for the Chinese people in negotiation and competition with the top-down propaganda regime.[8] The mainstream "state-over-society" view on contemporary Chinese nationalism cannot properly grasp the reality that the increasingly popular use of the Internet makes it more and more difficult for the state apparatus to monopolize the production and circulation of Chinese nationalist discourse, and that the Chinese people play a more active and independent role in the expression of "unofficial" nationalism around cyberspace. Even if the state intends to impose its will on the people, the public opinions on the Internet can never be effectively silenced and suppressed by the regime. On the Internet, the agent of Chinese nationalism is no longer the state or the party; the cybernationalism contains variegated popular imaginations that may support, oppose, or negotiate with the state's claim to legitimacy. In view of some local observers, the popular opinion on the net has formed a huge force in Chinese society, and expressed a "public demand for discursive right."[9]

Since the popular nationalism on the Internet is still in the making and its orientation is not fully determined yet, it might be too premature at this stage to make any claim as to what kind of impact the Chinese cybernationalism would have upon the state power, and how the state might react to the rising popular opinion from the grassroots.[10] To sustain its ruling authority, the Chinese government is still making every effort to promote the official version of patriotic education, while it is now losing its total control over the production, circulation, and consumption of nationalist discourse in the age of multimedia communication. Despite the persistence of state censorship, the rapid spread and use of the Internet have given the Chinese people new avenues of expression, and helped facilitate the development of civil society.[11] In the formation of the new public sphere, the Internet enables the Chinese people to experiment with their public opinions and to construct a dialogic space of communication.

In the case of anti-Japan protest, the Internet not only makes possible a platform for expressing nationalist sentiments, but also serves to formulate and organize protest action. Concerning the latter, in recent years, there have been some studies dealing with the rise of the Internet-fueled popular protests in China.[12] Most of these studies, however, do not place the rise of cybernationalism at the center of critical discussion. On the other hand, it seems that the recent comments on the issue[13] tend to conceptualize the Internet nationalism as parochial and irrational, while failing to capture the democratic potential of the self-organized protest action around cyberspace. What is presented on the net is not only anti-Japanese rage and hatred, but also a deliberative discussion over people's collective action.

The Discourse of Action: Debating Boycott Strategies

What is at stake here is the use of the Internet to discuss, debate, and formulate possible movement strategies for the anti-Japan struggle. This may be called a dialogic orientation of the anti-Japan action, which informs a cluster of vibrant political discourses and a pluralism of views. During the period of March and April 2005, the online campaign for "boycotting Japanese goods" (*dizhi rihuo*) was seen by many forum participants as a hopeful strategy to influence Japan's current attitude toward its wartime history. The proponents argued that Japan's economy has been closely tied with that of China for years, so the boycott of Japanese merchandise could hopefully put certain pressure on Japan over the history dispute. The evolvement of the online discussion and debate over the boycott action shows the popular democratic potential of the use of the Internet in China. That is to say, unlike the mainstream mass media manipulated by the state and elite groups, the new media has the potential to be used by movement activists as a public sphere,[14] through which the civilian participants/organizers of a mass protest campaign can get together to work out a plan and perhaps to reach a common ground on their long-term strategy. Such an Internet activism is likely to be a bottom-up, grassroots-oriented, and counter-hegemonic cultural practice cutting across the matrix of power and empowering civil disobedience.

In March and April, the boycott debate on the China-Japan forum could be divided into two major camps[15]: for or against the boycott. For the advocates, the campaign could effectively lead Japan to turn around its attitude with regard to the history of Japanese imperialist aggression. For those who opposed or questioned the merit of such a protest, however, if the boycott would have any impact on Japan's economic health, China's current economic development would be the first casualty, since China's economy has been allegedly inseparable from that of Japan.

What should be noted is that the forum does not seek to fix or determine the final meaning of the boycott campaign, but rather provides a site for the presentation of different discursive positions. The forum is used as a site for collective negotiation, contestation, and competition among different views. Cyberspace allows the forum speakers to freely share their views with each other, where the participants are quite conscious of the need to justify the cause of action. Here are some entries from the pro-boycott groupings, which may illustrate various ways of justifying the protest on the forum:

> Buying Japanese products is just paying tax to the Japanese government. Boycotting Japanese goods will reduce Japan's military expenditure and will make China safer. Boycott is patriotic.

> As our country's citizens, we have freedom to choose any product we want. Boycotting Japanese goods is a citizen's right to choose for himself. This is a human right of the majority of people who appropriate such a right to love their own country (just look at what Koreans have done).

> The government should take the lead to boycott Japanese-made cars, stationery and office supplies.

> We should never tolerate those Taiwanese businessmen who support Taiwan independence while making huge money in China. We should also object to the Japanese enterprises which make a profit in China but engage in anti-China activity.

> Boycott is Chinese people's responsibility and right. A more important factor is because purchasing Japanese merchandise is just like paying the military expenditure for the Japanese troops which ever killed 35 million Chinese civilians. This is not responsible to the victims and the heroes who died in the War of Resistance against Japanese aggression.

> Japan is a pragmatic country. If Japan doesn't get hurt, it would never turn around its attitude.

In addition to the narratives that defend and encourage the boycott of Japanese goods, there were also other voices calling for buying goods produced by Chinese companies. For the pro-boycotters, the practice of "patriotic consumption" is expected to have a great influence upon Japan's attitude and action with regard to its history of aggression, and may at the same time enhance the country's economic strength in competition with Japan:

> What needs to be done is not only boycotting Japanese merchandise, but also buying more national products!

> Perhaps our national products are not as good as that of other countries like Japan! But as Chinese, we should be more supportive to the goods made by our own country!

> I admit that many Japanese products' quality is much better than that of China's, but the quality is not that good enough for the Chinese people to give away the sense of dignity.

> The voice of boycotting Japanese goods is now challenging our national industries! . . . In the face of such a voice, I think it is now a good time for our national industries to reinforce their own sense of self-confidence and to speed up the process of gathering international market information to develop more products in return for our lovely compatriots! . . . Don't let us down, our national industries! Otherwise we would feel that the road of boycotting Japanese products is too long.

In the formation of the attitude, consensus, and collective action with regard to the boycott issue, the extremely lively web-based discussion and debate play a crucial role in drawing together a wide range of competing facts, opinions, standpoints, and emotional feelings expressed by the interested forum participants. The debate enhances the production and circulation of public discourse, which joins strangers and enables larger collectivities to make informed choices about their common future. Against the boycott campaign, the other group of forum participants, many of whom are as nationalistic and patriotic as their counterparts, then opposed the course of action with grounded reasons. For them, since China's economy has been deeply intertwined with that of Japan, the rejection of Japanese goods would only damage China's current economic development, leading to job cuts in the Japanese companies and factories in China. The following quotations are some illustrative examples of this kind:

> Nowadays, there is a surge of anti-Japan trends that encourage the boycott of Japanese goods and the support for Chinese goods. These conducts appear to be very emotional. From an economic point of view, these behaviors do not fit with the principle of maximizing comparative interest, since such behaviors would never help advance our economic and technological capability, but rather limit us from learning foreign advanced stuff, technology and product. Capital does not have political property; it can be used by us. If we exclude foreigners blindly, we would not be able to help our country but hinder ourselves from moving ahead.

> [The boycott] will badly damage China's market economic order established just now, resulting in a series of consequences such as bringing down the quality of the environment for investment and reducing the investment in China. It will not effectively smash the Japanese right-wingers, but will rather become more of a hindrance than a help, leading to withdrawal of investment from China (as desired by the Japanese right-wingers).

> I think that "boycotting Japanese goods" should be restrained on certain conditions. Since China has developed its economy, it is quite difficult to find a purely Japanese product on the Chinese market. Some likely "Japanese products" are made by our domestic companies with the new technology introduced from Japan, or are produced by a joint-venture company based in our country. We benefit a lot from these products. Should these products be included on the boycott list? What are we gonna do after boycotting these products? Who suffers most from the boycott? Especially in the circumstances of our country's entry into the international trade organizations, the door of our nation is open. What are we gonna do? . . . In the setting of trade liberalization, I think that the call for a boycott would only result in widespread reaction and negative impact on a global scale, which would push us to a disadvantageous situation. Those who suffer most would be us.

In addition to the polarized opinions over the action, there were also some entries calling for a more eclectic approach. One participant argued that the boycott should target those right-wing companies and the products they produce, instead of all products under Japanese brands indiscriminately. That is to say, as one entry wrote, the right-wing businessmen should be distinguished from "the good and virtuous Japanese people." Another discussant claimed that the ultimate goal of the boycott is not to promote extreme nationalism, but to facilitate China and Japan's friendship. Some postings then saw the boycott action as a short-term strategy, and urged Chinese companies to further their own competitiveness by learning and appropriating the advanced technology developed by the Japanese.

In my observation, each opinion camp represents a group of different opinions within a shared spectrum, and each group proposes a different set of views and strategies about the campaign. The popular movement is not simply a copy of the old-style "united front" which demands a homogeneous revolutionary unity without difference. Rather, the online debate creates a new fashion of solidarity, which is characterized by the synthesis of independent, autonomous, and unorganized participants. The above examples show that the online debate over the boycott action mixes together a diversity of views about China's development and modernization, comparative economic and political interest, as well as Japan and China's bilateral relations, and so on. It is the spirit of irreducible plurality, rather than homogeneous identification, which characterizes the spirit of the online discussion.

The diversity and complexity of the online opinion, according to a Chinese media critic, implies the maturity of Chinese cyberculture.[16] The collective participation in the discussion and debate over the boycott action also reveals the nascent form of Chinese "cyberdemocracy" (*wanglu minzhu*).[17] The online participants enjoyed an unprecedented high degree of freedom of expression. Moreover, in the shaping of popular nationalist politics, the Internet has become the backbone of communication for a broad coalition of individuals and groups. The extraordinarily open and unconstrained form of digital communication is particularly effective at establishing links among the forum participants (i.e., "horizontal" communication), providing a platform for sharing information and preparing further collective action. Such a political function is particularly significant for Chinese popular politics, since the media public sphere in China is largely controlled by the state agencies and elite groups in civil society, and the mobilization of popular protest is suppressed by authorities.

The above examples may provide a different way of knowing Chinese nationalism-from-below. First, it now seems necessary to pay attention to the dynamic process of Chinese popular nationalism, rather than simply condemn it from the state-centric view. Next, the existing Western liberal democratic theory's explanatory power is somewhat limited in the sense that it cannot properly describe and theorize the micro-level practice of the Chinese people and give such autonomous practices an adequate theoretical status. Chinese popular nationalism is too often dismissed and condemned as parochial, irrational, and chauvinistic, a "problem" that needs to be resolved or reformed in order to come up with the "modern" way of life.[18] Nowadays, however, it would be a grave mistake to make any claim about Chinese nationalism without paying attention to the lively debate on the net; in the China-Japan forum, there is not only an outpouring of indignation and anger, but also free expression and exchange of different views among the participants. Unfortunately, most existing research still assumes the "state-over-society" view and a parochial image about Chinese nationalism, failing to take into consideration the new democratic potential brought about by the cybernationalists. This being the case, what is needed is not simply dismissing or condemning Chinese nationalism, but rather a more sophisticated perspective and analytical tool to assess the ongoing political conjuncture in China.

Conclusion

The extraordinarily huge turnout of the anti-Japan protests is a forceful manifestation of the extremely powerful communicative and mobilizing potential of the new media for China's popular nationalism. This chapter tries to demonstrate some distinct ways in which the Internet is used by the people for contesting official discourse and mobilizing protest action in China. The illustration of online discursive struggles shows that the Internet serves as a key site for Chinese nationalism-from-below. With the rapid expansion of China's online network, the influence of online opinion on China's foreign affairs will inevitably grow larger and put more pressure on the state's claim to nationalist legitimacy in the near future. It is likely that the demand made by the cybernationalists will lead to a more sophisticated foreign policy-making process when dealing with China's neighboring countries like Japan, pushing the Chinese authorities further to take seriously popular opinion while promoting China's diplomatic relations. With the expansion of the virtual public space, the evolution and health of Chinese Internet culture will be significant not only on China's politics but also Northeast Asia's regional peace and prosperity.

The 2005 networked mobilization prepares the ground for a more dynamic protest cycle in China—Chinese cybernationalism has grown dramatically ever since. Despite the effort to constrain and at times suppress cyberprotests, the Chinese state has nevertheless become more responsive to the public, while seeking desperately to co-opt them into the regime, with calls through the state media for "rational" and "constructive" patriotism. In the post-2005 global context, China has witnessed several self-organized grassroots campaigns to reassert popular sovereignty and reclaim authority from below. These collective actions, which attempt to recover the right to rebellion and representation, signify a deep discontent of the present global hegemonic order and a desire to participate in global affairs. Chinese netizens have continued to speak out on various critical issues, urging the state to live up to its professed ideals, and pursuing a wider range of demands on the politics of recognition and national dignity. Given that the Internet remains the only public communicative space in China where popular sentiment can be expressed anonymously, most protest campaigns are launched online and spill out onto the streets unpredictably. These movements, aiming at not only rationales of foreign policy-making but also foreign governments' attitude and media coverage on Chinese-related affairs, are largely spontaneous and increasingly coordinated across national borders. The scenario of Olympic torch relay disruptions in Europe, and what Chinese see as biased and unbalanced news coverage of Tibet's unrest in 2008, provoked a new global wave of cyberprotests by ethnic Chinese around the world. They signed online petitions, staged public rallies, and formed "human walls" along the torch route via the Internet. After the disruption of the torch journey in Paris, for example, global ethnic Chinese netizens condemned the French government and circulated petitions to boycott French products from cosmetics group L'Oreal and the much-loved Louis Vuitton luxury brand. After rumors that executives of supermarket chain Carrefour supported Tibetan independence spread on the Internet in April, protesters gathered outside stores in several mainland cities calling for a boycott of its products. A "human flesh search engine" name-calling campaign was launched online with details of chosen targets being spread around discussion sites. Netizens were quick to discredit Western media, highlighting their blunders and accusing them of deep-rooted prejudice. These become the means for reasserting their rights and challenging global hegemonic publicity.

In addition to the alternative dialogic activity engaged in online, there also emerges a new kind of radical aesthetic practices that activate the democratic and participatory possibilities of netizens. With the growing popularity of weblog communication as well as video sharing and social networking sites among youngsters, the practice of cybernationalism has become more diverse, pervasive, and creative in producing subjectivity. The new sort of subversive communication becomes more interactive and collaborative, concentrating on the strategy of "culture jamming" that reappropriates official discourse or imagery and gives a totally different meaning. During the disruptions, Chinese netizens were active in uploading and sharing self-made videos, searching alternative sources of information on the situation, launching a global campaign against Western media outlets with the creation of www.anti-cnn.com, a website aimed at exposing allegedly biased coverage of China's handling of Tibet's unrest by Western media. Provocative counterspectacles were created to scandalize and subvert the dominant image, norm, and discourse of Western media or foreign adversaries. The campaign targeted not only the U.S. television network but all Western media outlets airing blunders and biased views in their coverage of the Tibet rioting. The anti-CNN website called on global Chinese to send protest letters, faxes, and e-mails to western media companies, urging them to apologize for false reports. The latest wave of demonstrations shows that Chinese cybernationalism has become more diffuse, flexible, and amorphous in terms of the power of self-organization, self-management, agenda-setting, and creative production—many of which are not simply an outburst of narrow-minded nationalistic discontents, but embody a desire for fairness and justice. These "voices without votes" signal a deep-seated crisis as well as a transformation of the moral and political legitimacy of the present global hegemonic order that fails to represent the wills and demands of citizens, and to obtain the consent of the governed.

Indeed, these transnational mobilizations, which take on some features of active sovereignty, have become a formidable force for checking global relations of domination from the grassroots level. In the age of cyberpolitics, not only state authority but also global hegemony is called into question. Rather than being effectively eliminated or suppressed by the state, these mobilizations have gained more support beyond national borders, reclaiming global citizen-subjects' right to representation and rebellion in various global affairs. These mobilizations have gained dramatic political force that is being translated into global communicative power affecting public opinion formation. The new transnational political public, however, does not always necessarily imply subverting the state's power. They also show that the global public opinion is no longer formed by a single, territorially bounded political citizenry or addressed to a single sovereign state. Instead, the discursive formation of the political will, which involves a wide range of nonnational interlocutors and communicative actions over digital networks, is addressed to the global imperial order and media hegemony. What makes the transnational mobilization distinctive is that the new publicity appeals its solidarity not as a "nation," but a global ethnic Chinese community. Although sometimes the transnational public may serve to promote a specific agenda tolerated by the state, their dynamics should not be conflated with the latter. The new development raises further theoretical and political questions that need to be examined critically: How do we perceive the way global netizens reclaim their "sovereign" authority and legitimacy against global hegemony, as opposed to the old state-centered framework and mode of thinking? What is the implication of such an active "sovereign" practice, and to what extent does it represent a sovereign principle and sovereign will different from the unified state-centric model? The subversive potentialities unleashed in the popular contestation have challenged a Westphalian frame which assumes a single, comprehensively bounded political space and identifies the public with the national citizenry of a sovereign territorial state. Whether Chinese cybernationalism would act as a counterweight or a complement to state power, and whether and how the global system could be made more accountable to the transnational political public, remains an empirical question for further research.

Notes

*The author would like to thank Simon Shen for his helpful editorial assistance.

1. Jim Yardley, *The New York Times* [query.nytimes.com/gst/abstract.html?res=F5 0E12FE39550C768 EDDAD 0894DD404482, accessed on 7 May 2005].

2. Liu Kang, *Globalization and Cultural Trends in China* (Honolulu: University of Hawaii Press, 2004); Yang Guobin, "The Internet and Civil Society in China: A Preliminary Assessment," *Journal of Contemporary China* 12 (2003): 453–75; Zhao Yuezhi, "Falun Gong, Identity, and the Struggle over Meaning Inside and Outside China," in *Contesting Media Power*, ed. Nick Couldry and James Curran (Lanham, Md.: Rowman & Littlefield, 2004)

3. Many Chinese websites, which contained protest-related information, were closed down by the Chinese authorities, with the most popular instant messaging client QQ taking measures to add the word "march" (*youxing*) to the list of banned words, preventing its users from sending messages that contained prohibited words (*China News Asia* 2005; *The Chosun Libo* 2005).

4. "Diaocha: Hulianwang Chengwei Souji Minyi Xintong" (Survey Internet Becomes a new channel of gathering public opinion), Xinhua Net [news.xinhuanet.com/it/2004-03/08/content_1350734.htm, accessed on 4 November 2009].

5. The year 2003 was declared by Chinese "netizens" (*wangmin*) as a "new chapter" of Chinese nationalism, with particular reference to the significant growth of web-based nationalist protest in that year. The form of these internet protests ranges from a signature campaign to online debate over such Japan-related issues as the Diaoyu Islands dispute, Japanese Prime Minister Koizumi's visit to the Yasukuni Shrine, the Japanese history textbook controversy, the Qiqihar Mustard Gas Incident, the Beijing-Shanghai high-speed rail link contract, the Zhuhai Prostitution Incident, and the Xibei University Incident.

6. The translation of all following quotations from Chinese is entirely mine.

7. Howard W. French, "By Playing at 'Rage,' China Dramatizes its Rise," *The New York Times*, April 22, 2005; Orville Schell, "China's victimization syndrome," *Taipei Times*, April 25, 2005.

8. Giese Karsten, "Speaker's Corner or Virtual Panopticon: Discursive Construction of Chinese Identities Online," in *Cyber China*, ed. Francoise Mengin (New York: Palgrave Macmillan, 2004).

9. Bin Shu and Chonglang Wang, "Zhongdian Xinwen Wangzhan: Xiangshang Shengzhang de Liliang" (Key news websites: the power that grows upward), *People's Net*, 2004.

10. Mujin Lee, "Wangluo Minzu Zhuyi Xiankai Zhongguo Minzu Zhuyi Xinpianzhang" (Internet nationalism starts a new chapter for Chinese nationalism), *International Herald Leader*, September 23, 2003.

11. Yang Guobin, "The Internet and Civil Society in China: A Preliminary Assessment," *Journal of Contemporary China* 36 (2003): 453; Qiang Xiao, "Cyber Speech," *Harvard International Review* (Summer 2003): 70.

12. Yuezhi Zhao, "Falun Gong, Identity, and the Struggle over Meaning Inside and Outside China," in *Contesting Media Power*, ed. Nick Couldry and James Curran (Lanham, Md.: Rowman & Littlefield, 2004).

13. Alan R. Kluver, "New Media and the End of Nationalism: China and the US in a War of Words," *Mots Plurieals*, no. 18 (2001), www.arts.uwa.edu.au/MotsPluriels/MP1801ak.html; Shanthi Kalathil, "Nationalism on the Net," *The Asian Wall Street Journal*, February 22, 2002; Chao Zhang, "Zhongguo de Jiduan Minzh Zhuyi Sichao Fenxi" (Analysis of China's Extreme Nationalist Thought), *Dangdai Zhonggguo Yanjiu* (Modern China Studies) 86, no. 3 (2004), www.chinayj.net/StubArticle.asp?issue=040303&total=86.

14. For a recent account of China's online public sphere, see *Beijing Review*, "Special Issue on China's Online Public Sphere" [www.bjreview.com.cn/ml-zhong Index-2005-34-z.htm, accessed on 1 September 2005].

15. This is not to suggest that there were only two streams of opinion on the issue. Here, I refer to the "polarizing" tendencies on the forum, especially during the last few days of March (when the street protest was about to take shape). On the polarization of online opinion, see Xiaodong Wang, comment on "Hulianwang yu Zhongguo Sixiang de Duoyuanhua" (The Internet and the diversification of Chinese thoughts), *BBC Chinese Net*, comment posted May 2, 2000 [news.bbc.co.uk/hi/chinese/china_news/newsid_742000/7422392.stm, accessed on 4 November 2009].

16. Xiaodong Wang, "Hulianwang yu Zhongguo Sixiang de Duoyuanhua" (The Internet and the diversification of Chinese thoughts), *BBC Chinese Net*, May 2, 2000.

17. Feng Liang, "Wangluo Minzu Weili Qiangda, Jingdong Zhongnanhai Jueceren" (The power of cyberdemocracy disturbs Zhongnanhai's policy-makers), *Asian Times*, April 15, 2005.

18. Jilin Xu, "Intellectual Currents behind Contemporary Chinese Nationalism," in *Exploring Nationalisms of China*, ed. George Wei and Xiaoyuan Liu (London: Greenwood Press, 2002); Chao Zhang, "Zhongguo de Jiduan Minzh Zhuyi Sichao Fenxi" (Analysis of China's Extreme Nationalist Thought), www.chinayj.net/StubArticle.asp?issue=040303&total=86.

5

Alternative Online Chinese Nationalism

Response to the Anti-Japanese Campaign in China on Hong Kong's Internet

Simon Shen

In April 2005, a series of anti-Japanese demonstrations broke out in China and South Korea to protest against "Japanese militarism" and Tokyo's bid for permanent membership in the United Nations Security Council (UNSC). Shih-Diing Liu published a perceptive article in *Inter-Asia Cultural Studies* in November 2006, describing these events as a milestone in the development of China's "online nationalism."[1] However, is this form of Chinese nationalism to be found throughout China and/or the Greater China Region? Do the mainland Chinese Internet users described in Liu's article—"Who we will call 'cybermen'?"—have a different nationalist attitude than their compatriots living on China's periphery in places such as Hong Kong? If Liu's observation were correct and representative of contemporary Chinese politics, would it be applicable for an eventually democratized China of the future?

Based on Liu's analytical framework, this chapter attempts to do the following in three parts: (a) study the concurrent and subsequent response of Hong Kong's cybermen to the anti-Japanese discourse prevalent within the mainland Internet community from April 2005 to March 2007; (b) analyze the reasons behind the differences; and (c) preview a likely pattern of online Chinese nationalism should political circumstances in Beijing one day more approximate those of present-day Hong Kong. To parallel major mainland Internet sources used by Liu, such as the Strong Nation Forum (SNF, *Qiangguo luntan*), Internet forums focusing on political discussion established in

Hong Kong will be our primary sources of reference. Major forums studied include the following:

- The Yahoo! Hong Kong Forum (YHKF)[2]: An extension of The Yahoo! Hong Kong, ranked as the most-visited forum in Hong Kong by Alexa Internet's statistics. Yet political discussion is relatively unheated.[3]
- The Hong Kong Discussion Forum (HKDF)[4]: Founded in 2004, claims to have two million registered members. It ranks as the second most-visited website in Hong Kong.
- The Uwants Forum (UF)[5]: Founded in 2003, managed by the Uwants Company Limited and consists of two hundred thematic forums. It claims to have over 1,200,000 registered members, ranking it as the third most-visited forum in Hong Kong.
- The Xocat II Forum (XIIF)[6]: Specializes in discussion relating to social science subjects in general.
- The Hong Kong Golden Forum (HKGF)[7]: Popular among the youth when it relates to computer hardware and software, generally seen as pro-democratic in Hong Kong.
- Japanese Knowledge Forum (JKF/Zhiri Bowu)[8]: A forum supported by the Chinese University of Hong Kong Student Union, and the students and professors in its Department of Japanese Studies. It aims to promote Japanese culture in Hong Kong and to exchange opinions in Japanese culture and politics.
- The History Forum KTzone (HFKTZ)[9]: A forum focusing on history.
- The Diaoyu Islands Defending Forum (DIDF)[10]: Run by the Action Committee for Defending the Diaoyu Islands (ACDDI), whose members are mainly activists.
- The Leftwing Forum (LF)[11]: A forum established by a local student named Li Roufou whose position it is to support the legacy of the New Leftists on the mainland.

Anti-Japanese Nationalism Online

Comparison between Mainland China and Hong Kong

In 2005, when mainland nationalists launched a series of physical demonstrations against Japan, "crowds" also gathered at Victoria Park, Hong Kong's trademark of protest, and marched to government headquarters. But it was reported that only about five hundred people participated in the protest, a level of disapproval similar to that shown for the Japanese bid for the UN Security Council permanent membership, a rightist version of a Japanese history textbook, Japanese prime minister Junichiro Koizumi's visit to Yasukuni Shrine, Tokyo's sovereignty claim over the disputed Diaoyu (Senkoku) Islands, and so on.[12] In comparison with the protest on the mainland, Hong Kong was entirely peaceful and there were no reports of damage to Japanese supermarkets or cars. As emphasized by one of the organizers, Albert Ho, vice-chairman of the Action Committee for Defending the Diaoyu Islands (ACDDI), "we shall not shatter glass or overturn cars, but we shall express our deepest sense of anger against Japan."[13] Unlike other larger-scale demonstrations organized in Hong Kong in recent years, such as the notable July 1 demonstrations in 2003 and 2004, the anti-Japanese demonstration was far more formality-oriented with relatively little passion shown among the crowd. Similar responses toward anti-Japanese Chinese nationalism in Hong Kong are also found in the local Internet community, which has never been as fervent as its mainland counterparts when foreign affairs are the topic of discussion. The differences between Hong Kong and mainland forums in response to anti-Japanese nationalism have three well-defined features and are described as follows.

Rational Cynicism Against Mainland *Fenqings*

To start with, the rationale for anti-Japanese Chinese nationalism was, in principle, shared by Hong Kong cybermen. However, most Hong Kong cybermen—regardless of their sentiments toward Japan—saw themselves as rational critics, but labeled the fervent anti-Japanese mainland nationalists as *fenqings* (furious teens).[14] Some users in Hong Kong, exemplified by the following posted in the JKF, juxtaposed *fenqings* with disdained Japanese rightists:

> Fenqings and Japanese rightists are basically the same: nationalist, and subjective. They share the same mentality, except in different positions. . . . The kind of people who are unable to think are the same in any place, aren't they?[15]

The appearance of this sort of argument in Hong Kong forums was significantly more frequent than on the mainland. From the logic they demonstrated, many Hong Kong cybermen identify themselves as Chinese nationalists and in registering mild protests against the Japanese cannot be simply described as pro-Western liberals. But at the same time, they were desperate to exhibit their disapproval of extreme anti-Japanese behavior in China. As one suggested in the HKGF, Hong Kong patriots were not coming from the SNF, "whose cybermen are all but fenqings."[16]

Such cynicism ran parallel with rationality in the Hong Kong cyberworld. Many messages in the SNF and other mainland forums were not only expressional but also functional, as they served the purpose of mobilizing participants to join the anti-Japanese demonstration or signature campaign. For instance, messages similar to the following were repeatedly found in the SNF:

> Kofi Annan has already pledged his support to the Japanese application for UNSC Permanent Membership. I am inviting my fellow countrymen to sign our names to express our opposition online. . . . For those who do not want to sign, please forward this to your friends.[17]

In fact, the entire anti-Japanese demonstration in Shenzhen and Beijing was "cyber-rooted."[18] This mode of online mobilization is by no means unfamiliar to Hong Kong cybermen, whose contribution in mobilizing participants for the July 1 demonstrations from 2005 to 2007 is well noted. Even on and before 1 July 2007, when the demonstration had waned considerably in Hong Kong compared with that of 2005, various messages were still posted mobilizing fellow users to join the protest.[19] However, there was almost no effort made in the Hong Kong cyberworld to mobilize other users to join the anti-Japanese demonstration, suggesting most Hong Kong cybermen were disinclined to organize massive nationalist movements themselves. Instead, a cool rational calculation of the effectiveness of any anti-Japanese movement was commonly found. Even nonviolent and politically correct movements organized by fellow Hong Kong citizens, such as those aimed at signalling Chinese sovereignty over the Diaoyu Islands, were dismissed as representing "extreme nationalism." A representative use of this calm and rational approach is seen in a posting titled "fenqings are as pitiful as Japanese youngsters":

> The protestors for the Diaoyu Islands from Hong Kong and Taiwan are patriotic. However, have they ever thought about the effectiveness of such protests? Screaming and shouting cannot force the Japanese government to relinquish their claim over the Diaoyu Islands. They'd better think of more practical means. Nationalism is not extremism![20]

Understanding of the Japanese Arguments

Another feature of the Hong Kong cyberworld's response to anti-Japanese Chinese nationalism that was largely absent on the mainland was the high degree of understanding of the Japanese arguments, no matter whether the arguments were being endorsed or not. For instance, Koizumi's visit to Yasukuni Shrine, which received constant criticism from Chinese leaders like Premier Wen Jiabao,[21] as well as mainland cybermen, went almost unnoticed by the Hong Kong cyberworld.[22] In the SNF, most cybermen strongly disagreed with categorizing the Yasukuni Shrine issue as a religious issue, and regarded the worship as a cult owing to the presence of souls of "class A" war criminals like Hideki Tojo.[23] Yet many Hong Kong cybermen who expressed a slight degree of anti-Japanese sentiment found the visit to be a "normal activity" of the Japanese, and chose to dismiss its political implications:

> I am not speaking on behalf of the Japanese. But actually, paying respect to war heroes' souls is understandable. Whether they are top-level war criminals is not a problem. The only problem is the social status of the visitor, Koizumi. If Koizumi is an ordinary citizen, then everyone would accept it, right? If so, why bother?[24]

In addition, some cybermen in Hong Kong seemed to be able to gather information released by the Japanese. For instance, a user of the HFKTZ simply treated those being worshipped in Yasukuni Shrine as martyrs:

> Besides the eleven top-ranking war criminals from WWII, Yasukuni Jinjya [Shrine] hosts many heroes and martyrs from the founding stages of modern Japan. Koizumi's visit may not represent his agreement with the war criminals. It may only represent his respect for the founding heroes of Japan.[25]

Another typical example showing Hong Kong cybermen's general understanding of the Japanese argument—and subsequently their distance from the mainland Chinese—took place on 23 June 2006. On that day, two Chinese men reportedly attempted to attack a Japanese policeman in Tokyo, resulting in the death of one and the arrest of the other. In response, China's Ministry of Foreign Affairs asked for a full investigative report, and mainland forums were again crowded with anti-Japanese messages. In contrast, most Hong Kong cybermen saw this as a minor issue that deserved no official diplomatic response. Quite a number of cybermen, such as the author of the following post on the JKF, even accused Beijing of stirring up trouble, as if it was only a local incident involving no foreigners:

> Can such minor stuff be regarded as a Sino-Japanese diplomatic issue? The Chinese government is just so tense. Why doesn't it simply attempt to govern the Chinese mobs in Japan?[26]

In a similar vein, many Hong Kong cybermen wish to be thought of as experts on Japan: even the pro-Beijing LF feature many rational analyses such as the following:

> Abe Shinzo may repeat the deeds of Junichiro Koizumi to gain more support from the extreme rightists. Strangely, although there are few extreme rightists in Japan, their influence in the economic and political arenas is not small. We are just not sure whether the Japanese seniors in the politics and business sectors have strong passions for the rightist thoughts or not.[27]

Localization vs. Nationalization: Absence of Criticism Toward Beijing

Abundance of References Made to Hong Kong

Although mainland forums are still heavily censored, according to Liu's study, some extreme anti-Japanese opinions had gone far beyond the official line.[28] For instance, the online critics Zaopao.com compared the PRC regime directly with the weak late Qing dynasty:

> Now we are in a critical moment. We hope Beijing is not imitating the government of the late Qing dynasty. Weakness would encourage the arrogance of our enemy. . . . We are unhappy to witness a successful China being as weak as the late Qing in handling Sino-Japanese relations.[29]

This post is cross-referenced in many mainland forums, such as the China918 Forum, which was later renamed the Patriotic Alliance Website.[30] Since anti-Japanese protests in China were nominally illegal, some forty protestors were arrested by the police.[31] Not unexpectedly, the arrests were heavily condemned by mainland cybermen. Following that, on 22 April 2005, two mainland discussion forums hosted by the "Chinese Federation of the Diaoyu Islands Defenders," China918 and Guangdong918, were shut down by the government.[32]

Indeed, it is not uncommon for Chinese nationalists to criticize their government by speaking in a patriotic tone. Dating back to the Belgrade embassy bombing in 1999, many online messages on the mainland have explicitly targeted the regime for its weak response toward the United States.[33] Similar criticism of Beijing's Japan policy after 2005 has indeed helped channel anti-government sentiments as well.[34] In contrast, Hong Kong cybermen—who are usually more critical of Beijing than their mainland colleagues—expressed far less criticism of Beijing's Japan policy and took no advantage of the situation to advance their criticism of the central government, not even the like-minded Hong Kong new leftists on the LF.[35] The focus of nationalists in Hong Kong who had shown anger toward Japan in local forums was rarely directed at the Ministry of Foreign Affairs, perhaps partly owing to the "one country, two systems" principle, which allowed no diplomatic and national defense issues to be handled in Hong Kong. Lacking an inclination to take part in national diplomacy, Hong Kong citizens criticized Beijing's official foreign policy, not surprisingly, somewhat mutedly.

One of the only exceptions to this in Hong Kong in recent years was found in an article that questioned Beijing's concessive attitude to Russian territorial claims. This was titled "When the Kuril Islands Become Sakhalin Islands," which was published in the mainstream media platform *Apple Daily* as well as on the author Lin Ji's Internet blog diary.[36] Yet the article received little attention from either the press or from the Internet, in sharp contrast to the online movies and commentary articles on local politics produced by the same author. He has a record of receiving tens of thousands of hits for one of his entries ridiculing the Hong Kong government's celebration of its tenth anniversary.[37] More importantly, the focus of the article—criticizing Hong Kong Observatory for its use of the Russian name—still targeted the HKSARG only, instead of Beijing.

As a mark of difference, it is unlikely that mainland cybermen would link Sino-Japanese relations to provincial politics in Guangdong or Shanghai. Yet when Hong Kong cybermen attempted to distance themselves from the anti-Japanese *fenqings* on the mainland, instead of commenting further on Chinese politics or Chinese diplomacy, they were more inclined to use the differences to talk about local politics in Hong Kong. In particular, many Hong Kong cybermen made reference to the official promotion of patriotic education as an attempt at decolonization of the special administration region (SAR): "the increasingly intensive anti-Japanese sentiment mobilized in Hong Kong is mainly owing to patriotic education promoted after the handover in 1997."[38] In the radically pro-democratic HKGF, many cybermen tied the anti-Japanese demonstrations in China to the growing anti-Japanese sentiments featured in Hong Kong's history syllabus:

> Questions in the history examination for college entrance only emphasize facts. Students trained in this system would focus on Japanese war crimes of the past, rather than Sino-Japanese cooperation of the future. National humiliation in the past should not be forgotten, of course, but we need to face the future.[39]

Limitations of Collection Sources

Before explaining the differences found between mainland and Hong Kong Chinese cybermen's responses to anti-Japanese nationalism since 2005, we need to acknowledge the limitations of the collection sources for this article. First, theoretically, is the establishment of the Internet aimed at promoting the free flow of information without geographical boundaries. We can only distinguish mainland and Hong Kong forums by two features: their different domain names and the different modes of Chinese used. In the Internet community, there are two major Chinese-character input systems: traditional Chinese (coded in "Big5") used in Hong Kong and Taiwan, and simplified Chinese (coded in "GB2312") used on the mainland. However, it is possible for Hong Kong cybermen to post messages on mainland forums by using Simplified Chinese, and the other way around. The HKLF, for example, is famous for the presence of mainland-originated members. It is technically impossible to completely separate the forums from one another, even though a differentiation between the two regions is obvious at present.

Second, on the mainland, the party's Propaganda Department (PPD) is responsible for containing anti-Japanese demonstrations, both physical and virtual, to a controllable level.[40] On 17 April 2005, the *Jiefang Daily* propagandized the importance of preserving stability, calling protesters to act "calmly and reasonably" (*lengjing lizhi*).[41] Two days later, Foreign Minister Li Zhaoxing told the PPD that "the masses . . . must believe in the Party and the government's ability to properly handle all issues relating to Sino-Japanese relations."[42] In response to Li's instructions, the PPD issued new guidelines for websites and media, resulting in the disappearance of most incendiary articles about Japan from mainstream online forums, effectively banning the latter. In contrast, online censorship basically does not exist in Hong Kong. It might, therefore, be logical to assume there is an even greater contrast between the two online environments while censorship is lifted from the mainland one day.

Lastly, owing to the fact that Sino-Japanese relations have not become a very hot topic among the Hong Kong Internet community, we have to reference ten different forums to remap the mosaics. For instance, anti-Japanese nationalism has never been given focus in the Independent Media Hong Kong (*Xianggang Duli Meiti*), a left-wing interactive forum in Hong Kong famous for its tendency to mobilize Internet users to participate in local political issues.[43] In addition, it is possible for some Hong Kong cybermen to mention the topic in passing, perhaps as part of their discussion on pop culture, but their comments are nonetheless recorded by our research. In other words, while anti-Japanese sentiment on the mainland would likely be more powerful if censorship was lifted, it may already be overstated in Hong Kong.

Interpreting the Differences

In the *Journal of Contemporary China*, Yang Guobin once argued that with the popularization of Internet usage on the mainland, anti-Japanese ideology was likely to spread fast and wide. This would both mobilize the mainland public to become engaged in additional activities and also muster an anti-Japanese public across the border.[44] While Yang's former prophecy is probably correct, the latter is yet to happen in Hong Kong. There may be three major reasons for the different responses to anti-Japanese nationalism seen in the mainland and Hong Kong forums discussed above. The first two relate to anti-Japanese nationalism in Hong Kong in general, and the third—more central to the argument of this article—is about the relationship between anti-Japanese nationalism and its online format in particular. These explanations are important to China in the sense that they may provide a reference point for ascertaining anti-Japanese nationalism in China should full democracy ever be granted to the country, which will be discussed in the conclusion.

Reaction Toward Patriotic Education after the Handover

As expected, Chinese nationalism was never encouraged during Hong Kong's colonial period. National education in Hong Kong at that time featured a depoliticized curriculum, talking about Chinese culture and history only from the Anglo-centric approach. Hong Kong citizens were, however, also discouraged from full integration with London. As a result, local identity in Hong Kong has only recently gradually developed, particularly since the 1970s. After 1997, former Chief Executive Tung Chee-hwa explicitly voiced his concerns about an identity confusion in respect to regional and national contexts and suggested promoting "national education" (*guomin jiaoyu*) in Hong Kong. The Committee on the Promotion of Civic Education (CPCE) of the Home Affairs Bureau (HAB) is responsible for promoting national education in the SAR via a soft approach such as television programs and advertisements.[45] In 2004, the CPCE and the Commission on Youth jointly set up the Working Group on National Education (NEWG) to formulate strategies and plans to promote national education to the general public. The Education and Manpower Bureau (EMB) is responsible for designing a new curriculum for students. For instance, the topic of "national identity and Chinese culture" has been added to the subject of "general studies." As questioned by the HKGF user introduced in the last section, the new design of the history syllabus expects students to merely remember hard facts, which could end up increasing anti-Japanese sentiment among the next generation.

However, anti-Japanese nationalism was yet to be propagandized by the new education system for a series of reasons. First, unlike the mainland history curriculum, that of Hong Kong focused on British colonial rule more than the Japanese invasion of China, making anti-Japanese nationalism a less notable theme from the start. As one person said in the XIIF, "the tragedy perpetrated by the Japanese on the Chinese will not be easily acknowledged by the new generation of Hong Kong."[46] Second, according to a research survey, many frontline teachers educated in the colonial curriculum were strongly hesitant about promoting patriotism in schools. Conscious of avoiding brainwashing the new generation, they taught national education in a way that usually resulted in discouraging students to react to China's modern history nationalistically.[47] Third, despite the dual nationalist and democratic identity possessed by many local politicians, as previously discussed, national education—interpreted as a top-down attempt by the central government to indirectly influence Hong Kong—had been a target of attack by pro-democratic politicians.

This partly explains why anti-Japanese democratic politicians failed to produce a large pool of anti-Japanese young people among their supporters. Lastly, some attempts to promote national education, such as a short advertisement clip called "Our Home, Our Country," which uses the Chinese national anthem "The March of the Volunteers" as theme music, were considered old-fashioned, causing some youngsters to label anti-Japanese sentiment and patriotism unstylish. For instance, several online messages commented that the short advertisement was "hard to put up with."[48]

The Role of Japanese Culture in the Hong Kong Identity

While mainland Chinese remain rather resistant to Japanese culture (and are interestingly more receptive to Korean idols), Japanese culture in Hong Kong has since the 1970s, with increasing financial transactions and trade taking place between Hong Kong and Japan (as represented by the influx of Japanese movies, idols, music, video games, and all kinds of trendy imports), been well merged into the local scene.[49] More important, this process overlapped the stage in the 1970s and 1980s when local people were developing their sense of a Hong Kong identity. Even the Hong Kong activists who initiated a social movement against Japan's sovereign claim to the Diaoyu Islands in the 1970s were heavily influenced by the influx of Japanese culture at the time. As a result, Chinese nationalists in Hong Kong have always taken a more rational stance toward Japan, as their innate emotions about Japanese culture were completely different from those found on the mainland. It is thus very difficult for people in Hong Kong to entertain boycotting Japanese products, both practically and emotionally, in the way mainland cybermen have proposed. The psychological complexity of the Hong Kong Chinese can be observed from messages such as the following:

> Do you know the Japanese constitution? How much do you know about Japan? You know nothing about Japan, but you are now initiating an anti-Japanese protest. I am against your behavior and see it as a silly and ignorant decision. Could you please understand more about Japan before getting furious about her?![50]

The importance of the above message, which was aimed at mainland cybermen initiating anti-Japanese protests, is its subtle reference to the fact that one of the themes of Hong Kong's local culture was to complain about the mainland's ignorance, or misunderstanding, of their compatriots in Hong

Kong. The post implied that while there was a lack of understanding among mainland cybermen on Japan, from Hong Kong's perspective, a similar misunderstanding among the same people could also be applied to Hong Kong.

When the cultural aspect is taken into consideration, the stereotypical image of mainland anti-Japanese fenqings in Hong Kong becomes even more destructive. It can best be seen from the controversies aroused by mainland Chinese actresses Zhao Wei and Zhang Ziyi. Zhao positioning herself alongside the Japanese military flag as a fashion statement sparked a massive Internet protest, which ended up with a nationalist physically assaulting her at a concert.[51] Likewise, Zhang was seriously criticized by mainland cybermen as a *hanjian* (traitor to the Han ethnicity) because of her leading role in the movie *Memoirs of a Geisha*; geishas are mistakenly seen as low-ranking Japanese prostitutes in China. Zhang was asked to give up her Chinese citizenship, or to "verify" her loyalty to China by committing suicide.[52] Therefore, when more serious anti-Japanese demonstrations are launched on the mainland, Hong Kong cybermen tend to make immediate reference to these two widely circulated examples which represent the parochial judgment and misunderstanding of mainland Chinese about external cultures, and resort to self-restraint against actively echoing their mainland fellows' views. For instance, when the anti-Zhang online movement became the finale to the anti-Japanese movement in 2005, most Hong Kong cybermen commented on it negatively, calling it "blind nationalism," and found it totally unacceptable. This was because the so-called wrongdoings of Zhao and Zhang were considered quite permissible by Hong Kong's mainstream society:

> Have you ever seen "Memoirs of a Geisha"? I have. The movie talks about Japanese society in the Warring period, and describes humanitarianism in detail. I don't think it is promoting any feudalist thoughts. Zhang is starring in a real society and representing the lower-class. . . . It is only a social history movie; has it promoted any ideology? I cannot understand the reasons stated by the mainland extreme leftists against this film.[53]

Multiple Identities of Anti-Japanese Organizations in a Free Society

A number of nongovernmental organizations have been set up in Hong Kong, the aims of which are to demand compensation from Tokyo for the Japanese occupation of the territory from 1941 to 1945. There are also now anti-Japanese organizations developing on these grounds. The Hong Kong Reparation Association (HKRA) founded by pro-Beijing activists is a typical example.[54] It has, however, close ties with its maternal political organizations. Notably, the HKRA frequently co-organizes anti-Japanese protests with the pro-Beijing political party the Democratic Alliance for the Betterment of Hong Kong (DAB), which was proudly included in their website's front page. They tend to promote anti-Japanese nationalism to the public in Hong Kong only for their own political interests; their identities rest on their political stance, not their anti-Japanese take.

Another prominent anti-Japanese organization is the Action Committee for Defending the Diaoyu Islands (ACDDI)[55]; the ACDDI can be considered the HKRA from the other side of Hong Kong's political spectrum. When many HKRA members were seen as having dual membership with the DAB, the ACDDI was set up by pro-democratic politicians, many of whom belonged to the Democratic Party (DP), or the more radical League of Social Democrats (LSD).[56] For example, Albert Ho Chun-yan, vice-president of the ACDDI, is currently the secretary-general of the Hong Kong Alliance in Support of Patriotic Democratic Movements in China as well as the chairman of the Democratic Party (Hong Kong). Chan King-sing, an active member of the ACDDI who went to the Diaoyu Islands in October 2006, is the convener of a tiny political party called the Saving Force of Democratic Hong Kong (*Minzhu Jiugang Liliang*), mainly formed by exiled democrats from the mainland, and a committee member of the radical democratic party, LSD. When it comes to nationalist issues, mainstream pro-democratic politicians in Hong Kong indeed never go beyond Beijing's line. However, to the public, the anti-Japanese gestures of the ACDDI and its members are more like part of their social activist movements than anything else. Since organizing social movements is a major way for the democrats to gain social influence and challenge the establishment in Hong Kong, ironically, the anti-Japanese stance assumed by Hong Kong democrats is often taken at merely face value by the public.

In his paper, Liu repeatedly stresses the bottom-up, autonomous "network struggle" as being the trademark nature of his coined online Chinese nationalism. However, as previously indicated, such a bottom-up network was formed not only against Tokyo but also against Beijing—as a euphemist way to advance Chinese rights to discuss official censorship. In Hong Kong, where freedom of expression still exists, there is no need for politicians and cybermen to overuse anti-Japanese discourse for their own political purposes, or for the sake of activating civil society. Since Sino-Japanese relations have been viewed by the general public in Hong Kong as one of the many calculations used by politicians from all sides for gathering their own political momentum, those opting for an independent and autonomous civil discussion online have simply tended to avoid joining the anti-Japanese debate, no matter their personal views. And those interested in politics would simply be attracted by the more prominent policies, without need to speak out indirectly via an authorized nationalist tone. To a certain extent, the anti-Japanese politicians are already alienated from the public in Hong Kong and only manage to absorb new members from their other political identities and affiliations. The relatively aloof response of the Hong Kong cyberworld toward Sino-Japanese relations, contrasted with the concurrent hot debates on other local political topics online in Hong Kong, and the heated discussions on Japan in mainland forums, is simply an extension of the above.

Conclusion: Previewing Chinese Online Nationalism after Democracy

To summarize, we found the response of Hong Kong cybermen to anti-Japanese nationalism differed from that of their mainland counterparts in four ways. There was a general display of rational cynicism toward the labeled mainland *fenqings*; more understanding was expressed of the Japanese point of view; there was a relative absence of criticism of Beijing's official Japan policy; and more references were made to local politics in Hong Kong. These differences might be explained by the reaction in Hong Kong toward patriotic education promoted after the handover and the role played by Japanese culture in the Hong Kong identity. Yet, more important, anti-Japanese sentiments in Hong Kong are also diluted by the multiple identities of anti-Japanese organizations, whose members—having stronger identities in other political fronts in a free society—find it unattractive to exploit the anti-Japanese debate to voice their other opinions. The remaining question is: Would the same situation be observed in mainland China should freedom of speech and press one day mirror Hong Kong?

We have stated earlier that if online censorship were to be lifted in China solely on nationalism, fiercer anti-Japanese sentiments might be witnessed. However, if censorship were lifted for all topics in China, the answer is likely to be different. As a matter of fact, the current political structure in China makes the nationalist discourse one of the only platforms for ordinary citizens to use in order to apply the strategy of "kicking the ball from the side (*chabianqiu*)" if they wish to promote their freedom, interests, and identity. They, indeed, have no or little intention of pressing the party-state to act on their expressions. The subtlety of this has not gone entirely unnoticed. Shih-Diing Liu also remarks that the ordinary citizens in China are developing their own form of nationalism (*renmin minzuzhuyi*)—"an autonomous political domain that is independent of the state nationalism."[57] Ordinary people have no other option but to use the online forum. This freely accessible network provides an outlet for the public to discuss Chinese politics. By using extreme wording or postings, activists can garner public attention and package the political message in terms of nationalism, which seems to stand for the state.

Once the new generation of Chinese had been equipped with the art of using split identity in the nationalist discourse, this became their platform for carrying out public discursive rights. Christopher Hughes analyzed the discourse of various identified Chinese nationalists and found that most of them were "either not particularly interested in nationalism or [were] highly skeptical concerning its possibilities for solving the problems faced by the Chinese state."[58] Considering the fact that public discursive rights have generally been lacking in the People's Republic since its establishment, mainland cybermen have achieved much from taking part in the nationalist discourse.

Yet, should full democracy and civil rights be granted in China, online users would no longer need to use nationalist discourse to speak for their intrinsic values and basic rights of expression. At such a time, no matter whether Beijing opts for a stronger Japan policy or not, at least within the online community fervent anti-Japanese sentiment is likely to be toned down significantly. This may not happen to as unresponsive a degree as observed in the Hong Kong Internet forums today, but the extravagant atmosphere described in Liu's article is also unlikely to reoccur.

Notes

This chapter is a revised version of the essay originally published in *Intercultural Communication Studies*, Vol. XVII:3, 2008, 172–86.

1. Liu, Shih-Diing. "China's Popular Nationalism on the Internet: Report on the 2005 Anti-Japan Network Struggles." *Inter-Asia Cultural Studies*, Vol.7, No.1, 2006, 144–55.
2. The Yahoo! Hong Kong Forum Yahoo! [hk.message.yahoo.com; accessed on 15 December 2006].
3. Alexa rankings for discuss.com.hk [alexa.com/data/details/traffic_details, accessed on 12 July 2007]. Same for the quoted rankings of other forums in the following section.
4. The Hong Kong Discussion Forum [www.discuss.com.hk, accessed on 15 December 2006].
5. The Uwants Forum [forum.uwants.com, accessed on 15 December 2006].
6. The Xocat II Forum XocatII [forum.timway.com, accessed on 15 December 2006].
7. The Hong Kong Golden Forum [forum.hkgolden.com/default.asp, accessed on 15 December 2006].
8. The Japanese Knowledge Forum [www.cuhkacs.org/~benng/Bo-Blog/index.php?go=category_18, accessed on 15 December 2006].
9. The History Forum KTzone KTzone [www.ktzhk.com/forumdisplay.php?fid=77&sid=K45nL, accessed on 15 December 2006].

10. The Diaoyu Islands Defending Forum [forum.diaoyuislands.org, accessed on 20 February 2007].

11. The Leftwing Forum [leftwing.hk-dn.com/bbs/index.php, accessed on 20 August 2007].

12. "HK police expect peaceful march." *South China Morning Post*, 14 April 2005.

13. "Gang zhouri youxing fanri jiang tujing rizi baihuo" (Anti-Japanese protests will stop by Japanese-owned supermarkets). *Apply Daily*, 13 April 2005.

14. In Cantonese, the character *fen* can be pronounced to mean human waste.

15. Message ID351. (2005, November 13). Response to "Zhongguo fenqing ji riben youpai weixie yanlun ziyou" (Freedom of speech is abused by *Fenqings* and Japanese rightists). Retrieved 1 August 2007 from JKF.

16. Message ID894554. (2007, March 30). Response to "Qiangguo fenqing xilie fanre youxing" (*Fenqing* Action Series). Retrieved 5 August 2007 from HKGF.

17. Message ID4210. (2006, April 15). "Fandui riben jiaru lianheguo changren lishiguo" (Oppose Japan joining as the permanent member in the United Nations' Security Council). Retrieved 5 August 2007 from SNF.

18. "China's cyber-warriors on the march." *South China Morning Post*, 7 April 2005.

19. Message ID976441. (2007, July 1). "Nide hui jidian qu weiyuan canjia youxing?" (When will you join the July 1 March in Victoria Park?). Retrieved 10 August 2007 from HKGF.

20. Message ID4353112. (2007, July 13). Response to "Zhongguo fenqing yu riben xinshengdai tongyang kelian" (*Fenqings* are as pitiful as Japanese teenagers). Retrieved 10 August 2007 from UF.

21. On 3 April 2007, Wen again commented that the issue was something "deeply hurtful to Chinese emotions" and demanded the Japanese stop visiting the shrine.

22. "Wenjiabao: xiwang rifang jinhou tingzhi canbai jingguo shenshe" (Wen Jiabao: We hope the Japanese government will stop paying respects at Yasukuni Jinjya again). 4 April 2007. *China Review News*.

23. Message ID613611. (2006, July 4). Response to Yinghe Riben ren ba Jingquo Wenti Daoxiang Zongjiao qu Jieshi shi Yangai le Riben ren de Cuowu de Lishiquan, Jue Bu ke gu [Treating the Jinjya issue as a reglious issue to cater to Japan, is ignoring the mistaking perception of history which is not acceptable!]. Retrieved 12 August 2007 from SNF.

24. Message ID703472. (2006, September 20). Response to "Xiaoquancanbai shenshe" (Yasukuni Jinjya visit). Retrieved 10 August 2007 from HFKTZ.

25. Message ID703472.

26. Message ID69. (2007, March 22). Response to "Riben jingcha qiangsha zhongguo yifan jiang cheng waijiao fengbo" (Chinese shot by Japanese police). Retrieved 12 August 2007 from JKF.

27. Message ID8457. (2007, June 26). Response to "Riben houhui le ma?" (Has Japan regretted?) Retrieved 12 August 2007 from LF.

28. Liu, Shih-Diing. "China's Popular Nationalism on the Internet. Report on the 2005 Anti-Japan Network Struggles." *Inter-Asia Cultural Studies*, Vol.7, No.1, 2006, 144–55.

29. Message ID903. (2007, July 9). "Xiwang Zhongguo zhengfu buyao chengwei wanqing Zhengfu" (Hope Beijing is not the government in late Qing dynasty). Zaopao. Retrieved 10 August 2007.

30. Aiguozhe Tongmeng Wang (The Patriotic Alliance Website). [x.1931-9-18.org, accessed on 20 August 2007].

31. "Shanghai ju 42 fanri shiweizhe" (Shanghai arrested 42 anti-Japanese protesters). *Mingpao*, 29 August 2007.

32. "Zhongguo guanbi bufen fanri qangzhan" (Some anti-Japanese websites are shut down by Beijing). 22 April 2005. *BBC China News*. Retrieved 15 April 2007 from the Chinese Federation of the Diaoyu Islands Defenders (www.cffdd.org.cn); China918 (www.china918.cn/forum.asp); and Guangdong918 (www.gd918.org).

33. Sang, Wei. "Zayiza Mei Shiguan You Shenme Liaobuqi?" (What a Big Deal to Destroy the American Embassy?). *Mingpao Monthly*, August 1999, 52–55.

34. Message ID113817. (2006, August 7). Response to "Xiaoquan yangyan suishi canbai jingguo shenshe, ni zenyangkan?" (Junichiro Koizumi will visit Yasukuni Jinjya, what do you think?). Retrieved 12 August 2007 from SNF.

35. Message ID8507. (2007, May 6). "Riben baofa dayouxing yaoqiiu rizhong duanjiao!" (Protest broke out in Japan!). Retrieved 12 August 2007 from LF.

36. Lin Ji's Internet blog diary [plastichk.blogspot.com/, accessed on 20 August 2007].

37. Lin Ji's Internet blog diary.

38. Message ID8. (2006, August 12). The anti-23 discussion group. Retrieved 15 August 2007 from HKDF.

39. Message ID978601. (2007, July 7). Retrieved 15 August 2007 from HKGF.

40. Xu, Zhiyuan. *Aiguo Wuzui* (Innocent to be Patriotic). Hong Kong: TOM (Cup Magazine) Publishing Ltd, 2005; Liu, Guokai. "Zhongguo Wei Minzuzhuyu De Youyici Baolu" (Another Exposure of Chinese Pseudo Nationalism). *Liu Guokai Wenji* (Collected Works of Liu Guokai), 12 April 2001 [www.boxun.com/hero/liugk/5_1.shtml, accessed on 10 December 2005].

41. "Shanghai shi zhengfu fayanren jiu bufen xuesheng qunzhong duiri shiwei dawen" (Shanghai government answers media on anti-Japanese march yesterday). *Jiefang Daily*, 17 April 2005.

42. "Waijiao buzhang lizhaoxing jiu zhongri guanxi baogao" (Foreign Minister Li Zhaoxing reports on Sino-Japanese relations), Xinhua Net, 19 April 2005.

43. Hong Kong In-Media (*Xianggang Duli Meiti*). Retrieved 15 August 2007 from www.inmediahk.net/.

44. Yang, Xiao. "Rediscovering Republicanism in China: Beyond the Debate Between New Leftists and Liberals." *Contemporary Chinese Thought*, Vol.34, No.3, Spring 2003, 18–34.

45. "RoadShow Civic" Education Series. 20 December 2006.

46. Message ID167892. (2006, September 10). Retrieved 18 August 2007 from XIIF.

47. "12% xuexiao congwu shengqi xiguan" (12 percent of schools in Hong Kong have never raised the national flag). *Wenweipo*, 29 June 2007.

48. Message ID756019. (2006, October 9). Retrieved 18 August 2007 from HKGF.

49. Japan is Hong Kong's third-largest trading partner after mainland China and the United States, and Hong Kong is Japan's ninth largest trading partner. Bilateral trade in 2006 totaled 4.41 trillion yen, with Japan's exports accounting for 4.24 trillion yen and its imports for 177 billion yen.

50. Message ID8. (2006, May 26). Retrieved 18 August 2007 from HKDF.

51. "Zhaowei 28 riwen zai Changsha yanchu zao yiming nanzi xiji bei tuodao" (ZhaoWei was attacked on 28 December night show in Changsha). bbs.people.com .cn, 29 December 2001.

52. "Zhongri wangmin qiqi paohong zhangziyi" (Zhang Ziyi was condemned by Chinese and Japanese Internet users). *Singapore United Morning Post*, 30 November 2006.

53. Message ID4991. (2006, February 4). Response to "Woguo jinying dianying Yiji huiyilu" (My country banned the publication of *Memoirs of a Geisha* the movie). Retrieved 18 August 2007 from LEFT.

54. Hong Kong Reparation Association Website. Retrieved 15 March 2007 from home.netvigator.com/~hkra2002/index.html.

55. Action Committee for Defending the Diaoyu Islands Website. Retrieved 5 August 2007 from www.diaoyuislands.org.

56. Interestingly, members of the Civic Party, a new pro-democratic party in Hong Kong gaining increasing popularity, are trying to deviate further from the nationalist line and avoid declaring high-profile stances in Sino-Japanese relations.

57. Liu, Shih-Diing. "China's Popular Nationalism on the Internet. Report on the 2005 Anti-Japan Network Struggles." *Inter-Asia Cultural Studies*, Vol.7, No.1, 2006, 144–55.

58. Hughes, Christopher. "Interpreting Nationalist Texts." *Journal of Contemporary China*, Vol.14, No.43, May 2005, 247.

6

Ethnocentric Perception Re-explored

Online Chinese Nationalism Toward Taiwan

Benson Wai-kwok Wong

THE EMERGENCE AND RAPID POPULARITY of YouTube in 2005 has not only provided a unique arena for Internet surfers producing and uploading video and/or audio clips on the website, but has also facilitated dynamic and interactive communication among users through the sharing of these clips and the consequential comments left after these are shared. YouTube provides one of the most influential platforms, apart from forums and chat rooms, for surfers to freely express their opinions in verbal and visual form. Unlike such mainstream media as television, newspapers, and radio, in which media enterprises, serving as gatekeeper, select and highlight "representative" views of the public to release and disseminate, the boundlessness of the Internet has constituted a challenge to the mainstream media mainly dominated by the powerful, such as government officials, capitalists, and politicians, regardless of liberal governments or authoritarian regimes.

Referring to Chinese context, Internet use has flourished dramatically. A survey has reported over 210,000,000 surfers as of December 2007, representing 41.1 percent of the total surfers in the Asian region, with an increase of 833.3 percent compared to the figure in 2000.[1] However, the Chinese government has taken the dominant approach to regulate Internet use, and has imposed strict restrictions and taboos on it, including the proscription of such politically sensitive topics as the Tiananmen Square events of June 1989, the Falungong movement, and the exclusion of critical voices toward top leaders at central and local levels, as well as controversies surrounding poor governance and maladministration.[2] On the other hand, the official patronage of patriotism after the 1989 Tiananmen crisis stimulated the upsurge of Chinese nationalism in the form of upholding "reform and opening up" and "say no to the West,"[3] with the aims of relegitimizing hard authoritarian governance, upholding growth and development, and resisting the influence of liberalism coming from the West. Additionally, the controversies between China and the United States, as well as Japan, have attempted to exact historical revenge and promoted ideological conflict, facilitating antagonism and eventually promoting patriotism in the long run. Overall, the strengthening and flourishing of nationalism, the rejection of external intervention, and the ultimate unification with Hong Kong, Macau, and Taiwan are three unchallenged beliefs held by the Chinese authorities.

To explore the nature of online Chinese nationalism toward Taiwan, this chapter uses a video clip placed on YouTube as a means of unraveling different perceptions, attitudes, and responses to "the Taiwan issue." Through a textual interpretation of comments on the clip, this chapter outlines how online Chinese nationalism has been presented, narrated, and interpreted in the Internet setting. This chapter is divided into four parts. Part 1 is a conceptual and empirical examination of Chinese nationalism toward Taiwan. Part 2 introduces the method adopted to collect and analyze such comments on the video clip. Part 3 focuses on a textual analysis of the comments left by Internet surfers proposing relevant themes. In the concluding remark, this author synthesizes the above analysis by inferring the logic behind the presentation of such a political discourse.

Chinese Nationalism Toward Taiwan: Official and Online Versions

Before exploring online Chinese nationalism in the conceptual and empirical aspects, it is important to understand the official position of Beijing on Taiwan so as to serve as a foundation for understanding its nature. One of the core elements of the concept is that the Taiwan issue is only about domestic politics between the CCP and the Kuomintong (KMT), hence external forces should not interfere with the entire process of initiation, negotiation, integration, and unification between the two political parties.[4] Another element is that while the "One Country Two Systems" concept would be implemented so as to maintain the unique system on the island, Taiwan would not enjoy "absolute autonomy," since this implies the recognition of "two Chinas."[5]

> Beijing rejects the presence of "two Chinas," indicating that the Nationalist government in Taiwan is not a *de jure* government with sovereignty and legitimacy. This is not a negotiation between two governments, but between two political parties, with the aim of reinforcing the domesticization of the Taiwan issue and

allowing no space for intervention by any "external force" (Weilai Shili)—a term solely defined and interpreted by Beijing.

The rejection of "two Chinas" can also be found in the concept that when the PRC establishes diplomatic relations with another country, that country must terminate relations with Taiwan. Also, when Taipei opens diplomatic relations with another country, Beijing ceases its relations with that country so as to avoid the "two Chinas" situation.[6] In summary, the above assertions have basically shaped Chinese nationalism, and are supported and expressed in different domains, including in the cyber setting.

In retrospect, Taiwan shapes a salient significance in Chinese nationalism. After Hong Kong's and Macau's return to China in 1997 and 1999, respectively, Taiwan has become the last territory politically departed from the PRC. Under the mentality of completion (*Wancheng* or *Wanman*) prevailing in Chinese culture, Taiwan's continued separation leaves the dream of a finally reunified China incomplete and unfulfilled. Resolving the impasse remains a key objective of the Chinese leadership—an objective that has passed into the popular discourse resulting in the unease and restlessness of the government and the people. In short, as Susan L. Shirk argues, "the roots of the Chinese fixation on Taiwan are purely domestic, related to regime security, not national security."[7] Another salient reason is that the increasing divergence between Beijing and Taipei in its political systems, practices, and cultures further frustrates Beijing authorities, notably after the Taiwanese undertaking of democratization since the late 1980s, universal suffrage of the national leaders as well as the legislature since 1996, and the transfer of power to the Democratic Progressive Party (DPP) in 2000. While the PRC may realize that the political transformation on the island cannot be reversed and interrupted, the further widening of such a political gap between the two places would only exacerbate existing obstacles to the attainment of national unification. Above all, while Beijing authorities devalue the political transformation of Taiwan by highlighting the political "chaos" in the legislature, and the incompetence of the DPP in dealing with economic troubles, political transformation on the island reflects that the PRC worries about political and ideological differences between two places hinder national unification. In this regard, Chinese nationalism is one of the most powerful political tools with the focus of reminding and warning the Taiwanese not to forget their historical and cultural roots, not to drive toward and support independence, and not to resist national unification by any means, including the intervention of external forces.

Xu Wu has adopted the identical term "Chinese cybernationalism," which is defined as "a non-government sponsored ideology and movement that has originated, existed, and developed in China's online sphere over the past decade starting from 1994. It is a natural extension from both China's century-long nationalism movement, but it is different from both the Chinese Communist Party's official version of patriotism, and the traditional Chinese nationalism movement."[8] However, Xu does not elaborate on the details relating to the major differences between three types of nationalism in light of (1) nature, (2) discourse, and (3) impact in shaping local politics and foreign policies. Also, he seems to ignore the fact that with the strict control over the use of the Internet, surfers tend to be conformist in discussing domestic issues, and supportive in chanting the glory of the motherland, aggressive and hostile in expressing views on foreign countries, notably the United States and Japan,[9] and in criticizing the political practice and transformation of Taiwan, regardless of democratization and independence.

Undoubtedly, Chinese nationalism is such a complicated concept that it is difficult to concretize into a clear and succinct definition. Adopting mainland China's perspective, Xu (2006) has argued that Chinese nationalism can be understood in two aspects: (1) an exit strategy aimed at remedying the collapse of Communism and (2) a new identity intending to regain political legitimacy and credibility.[10] Most importantly, Xu highlights the importance of the Taiwan issue in shaping Chinese nationalism in the political, strategic, and diplomatic sense. Politically, the mentality of a united and centralized government is replete with references to historical experience. Strategically, Xu has used the American Civil War as a metaphor to describe the tense relations between China and Taiwan, noting that the PRC has adopted the flexible approach to deal with the unification issue. Diplomatically, both the United States and Japan are hindrances since they support the independence movement on the island.[11] Such an assertion is also shared by Peter Hays Gries, who argues that Chinese nationalism can be understood as a nationalistic interaction with other nations, notably the United States and Japan.[12] In this connection, "Taiwan has been a symbol and rallying point that unifies all spectrums of China's nationalism branches. The government-promoted patriotism, grassroots nationalism, and elite nationalism, all share the same level of intensity and the same kind of mentality toward the Taiwan independence issue."[13]

Reflecting Xu's assertion on Chinese nationalism toward Taiwan, it should be noted that Internet surfers not only firmly affirm the official lines toward Taiwan, such as the upholding of ultimate unification with mainland China, the opposition to Taiwan's independence and the external intervention, the persistence of the military means to liberate Taiwan, and the adoption of the "One Country Two Systems" to complete unification, but also tend to present their opinions in an assertive and even aggressive manner, with the occasional use of rude and foul language, which is delineated and discussed later.

While Xu argues that Chinese nationalism toward Taiwan can be manifested in the recovery of national sovereignty on Taiwan Island, he might be missing a key discussion in the ongoing contest between the Beijing and Taipei governments before 2000 regarding the claim of a legitimate Chinese government.[14] Xu might underestimate the importance of the Taiwanese authorities. During Chiang's era, the recovery of mainland China could have been seen as a political discourse, emphasizing that the Chinese government in Taipei did not give up its claim of being a legitimate Chinese government under the banner of the Republic of China (ROC). Of course, with the increasing localization of Taiwan's politics since the 1990s and the administration of the Democratic Progressive Party from 2000 to 2008, Taiwan's identity, as reflected in the promotion of "loving Taiwan," has proliferated. The withdrawal of unifying with China, in fact, has attracted severe criticism from surfers supporting the complete unification with mainland China since they think that the Taiwanese have recognized the *de jure* independence of Taiwan.

Methods

To locate, identify, and delineate the various opinions and perceptions of surfers in mainland China toward Taiwan, a video clip five minutes long was uploaded and made publicly available on YouTube at hk.youtube .com/watch?v=tN8aEQVOoN8. The clip had registered 65,768 views from September 2006 to May 2008. The author has likewise offered the function that allows surfers to leave comments about the clip. The video clip was extracted from a popular TV commentary program entitled *Xiwen Gagaga* produced by Super Television of Taiwan.[15] The content of the clip revolves around two hosts sharing their experiences and observations about mainland Chinese. They start with a narrative about how a mainland Chinese passenger poorly treats a Taiwanese flight stewardess by intentionally humiliating her because of her Taiwanese identity. This video clip actually serves as a stimulus with the objectives of (1) tapping the opinions of surfers on the clip, (2) identifying and categorizing various opinions, and (3) examining the ways by which viewers perceive and present themselves in relation to the clip and to Taiwan. In this study, the author focuses on a total of 996 comments left from September 2006 to May 2008 for data analysis. After skipping meaningless comments (e.g., just leaving a few words without any clear meaning), and messages with only foul language, those comments that presented themselves in a clear and meaningful manner were chosen. During the extraction of the comments, their usernames were also attached for identification. To maintain the originality of the meaning and language, all grammatical faults and errors were faithfully preserved in case they were written in English, with a bracket to clarify and revise further if necessary. The bracket was also used for filling in incomplete statements without changing the original viewpoint and meaning of the comments.

Interpreting Online Chinese Nationalism Toward Taiwan

To respond to and support the anti-independent stance, some of the forums adopt the banner of anti-independence or national strengthening to attract and solidify surfers expressing identical views that attack Taiwan's independence. One of the forums, entitled "Anti-independence of Taiwan" (*Fan Taidu Luntan*) at www.anti-td.com, provides an arena for collecting information and comments criticizing Taiwan's independence and advocating national unification. Another site, this time called a strong country forum (*Qiangguo luntan*) at bbs.people.com.cn under the *People's Daily*, also covers comments about China unifying with Taiwan and criticism of the ignorance of Taiwan in not wanting to be a part of China. Moreover, some of the mainland surfers either upload clips of official speeches by current and past national leaders, or respond to the clips found in YouTube relating to Taiwan's independence. For example, one of the video clips is about Zhu Rongji warning of Taiwan's independence.[16] A total of 207,496 surfers had seen the clip and there were already 1,454 comments as of January 2009. An interesting remark made by Zhu regarding the clip is that "in the case supporting Taiwan's independence, there will be no good ending."[17]

By focusing on the video clip introduced previously for textual analysis and then synthesizing and integrating different views, this study proposes the following three features of online Chinese nationalism toward Taiwan—namely, hostility, supremacy, and patriarchy.

Hostility

While most surfers do not comment on the content of the video clip, some of them have responded to issues negatively relating to the Taiwanese/Taiwan. Such hostility is formed by historical and contextual factors. After the KMT moved to Taiwan in 1949, the political division between mainland China and Taiwan became one of the key issues in recovering national integrity and in achieving national unification in the eyes of the PRC. With the passage of time and the widening gap of political difference between the two locations—Taiwan has been democratizing since the mid-1980s whereas China has maintained and consolidated hard authoritarianism—the vision of unification seems to be far from being fulfilled despite the rise of China due to economic reforms and its opening up since 1978. Hence, both sides may lack shared and mutual understanding in light of political, social, and cultural aspects. Contextually, the political transformation of Taiwan has antagonized some Chinese on the mainland, especially when they are puzzled by why the Nationalist government in the post-Chiang era no longer suppresses or bans the independence movement, an important political gesture to demonstrate that Taiwan is determined to recognize itself as a part of "China." When Chen Shui-bian won the presidential elections in 2000 and 2004, respectively, Chen's persistence of independence in the form of promoting localization in various areas led to severe criticism from the local pro-unification camp, the so-called deep-blue force, as well as from China. In this regard, some mainland surfers believe that the Taiwanese have no intention of uniting with mainland China, and doubt why the Taiwanese supported and even cast their votes for the DPP, a political party supporting Taiwan's independence. They believe that such a trend not only encourages Taiwan to become a *de facto* independent state ultimately, but also undermines the dream of national unification that shows the might of China. With respect to the above delineation, some mainland Chinese surfers expressed their hostility toward Taiwan by engaging in verbal attacks and threatening further action. For example,

- "The Taiwanese, the motherland will annex you sooner or later." (imhyy)
- "Why is the program so naïve? I have nothing to say. What they (hosts) said are bullshits. The mainlanders do not do that. It is a shame that they (hosts) tell the lie." (BeijingBoy1985)
- "I got nothing to say. . . . I don't even know how suck 'Taiwanese' is until today . . . watch yourself . . . we will revenge as soon as you laugh at us." (Chinaman089)
- "after 2008, (I/We will) kill all Taiwanese!" (Paradise2046)
- "Zheng Hong-yi (N.B. the male host) is 87 (that is 'Baici,' means stupid)." (smshyu0615)
- "Taiwan is a part of China / Anyone who wants to split it from China will get a bad end! / I think the one who always think(s) Taiwan (is) better than Mainland is just a kind of mental masturbation! / When you wake up from your mental masturbation, you'll find that Mainland preponderates over you / Over the Five Stars Red Flag is the legal national flag of China in this world! / PLA will liberate Taiwan!" (BeijingGirl1)
- "narrow-minded, biased, and ignorant island people." (davidimhan)
- "The male Taiwanese speak in a sissy manner." (zyp00)

In expressing the hostility mentioned above, the surfers usually adopted the following manner. First, through personal humiliation and an attack on the Taiwanese such that the surfers demonstrated their ferocity and violence. With the manifestation of the collective expression, the mass force produced a tremendous influence on the Internet setting, noting that they have shaped a dominant voice in responding to the divergent views of patriotism. Second, they posed a threat to the Taiwanese by repeating the assertion of "adopting military action to annex Taiwan so as to complete the unification." Both manners share the same goals: surfers should fully and unconditionally accept and recognize their views, and warn, rather than persuade with rationality and respect, other surfers not to leave similar opinions so as to establish and maintain a dominant position in the mainstream political discourse within the Internet environment. This can be associated with the manners by which the pro-Maoist supporters violently and furiously attacked those who were labeled as "anti-revolutionaries," "capitalists," or "imperialists" during the Cultural Revolution (1966–1976), who were regarded as "class enemies" and "traitors." This manifestation can also be found in Taiwan's independence, regardless of political belief or cultural affirmation, and may be attributed to the closer tie with such countries as the United States and Japan, as shown in the following messages:

- "With the increasingly might of the motherland, the ambition of the United States and Japan has been exposed, they would not care about the Taiwanese apart from defending their own interests. The Taiwanese, be awakening! Do not serve such Westerners to hurt your compatriots!" (bradenwang)
- "Taiwan only learns from the little Japan. The Taiwanese soap operas are copied from the Japanese comics. Such Japanese comics are not stylish at all. It seems that (Taiwan) has been colonized by the Japanese for such a long time so that (the Taiwanese) have no rational thinking. Without the economic help from mainland China, what can you do except saying the empty words?" (shuimuashuimu)

The above messages highlight the intentional deployment of the political and historical hatred against Americans and Japanese to justify the antagonism. In the eyes of such nationalist surfers, such a negative feeling is reasonable because the Taiwanese have not only forgotten about their ancestry but have also taken due notice of the "betterment" of the enemies, notably Japan. It is therefore reasonable to impose such a harsh criticism to remind the Taiwanese that they cannot historically, politically, and culturally deviate from China.

Supremacy

The case for supremacy can be perceived as a situation of political inequality and discrimination between the powerful and the powerless within a given setting, and the adoption of coercion to exercise control.[18] This has been virtually incorporated into the methods of expression by such surfers who tend to support the ultimate unification of Taiwan with China. Despite the fact that the terms like supremacy, hegemony, and manipulation are not usually adopted by PRC authorities to describe the relations with Taiwan, this does not imply that an equal footing and mutual understanding are the core elements featuring Taiwan under the PRC's political narrative. In fact, the term "China," regardless of political, territorial, ethnical, cultural, or historical sensibilities, should not be challenged and denied. China's supremacy over Taiwan can be examined in three ways: First, historical supremacy can be understood in the salient remark "Taiwan is a part of China," found in official statements and pro-governmental publications. The PRC authorities interpret the term "China" or "Chinese government" in an ambiguous manner despite the fact that the authorities intentionally neglect and mute the fact that either "China" or "Chinese government" is under the banner of the Republic of China on Taiwan. Let us refer to the *People's Daily*, which presents the typical view of the relationship between "China" and Taiwan.

> In December 1943, the Cairo Declaration jointly issued by China, the United States, and Britain stipulated that all the Chinese territories that Japan had stolen from China, including the island of Taiwan, be returned to China. The 1945 Potsdam Proclamation reiterated that demand.
>
> On October 25, 1945, Taiwan was returned to China. Taiwan's status as part of China has not changed since then, the article stressed.
>
> The Cairo Declaration and Potsdam Proclamation are both widely recognized as legally effective documents of international law. And, after the signing of the two documents, the United States and its allies declared its recognition of Chinese sovereignty over Taiwan on many occasions.[19]

Note that in this text, the PRC authorities used "China," "Chinese government," and "China sovereignty" interchangeably to represent itself; but in fact the political entity entering into international agreements and treaties at this period was in fact the Republic of China. Thus the term "China" contains a deliberate ambiguity of the political meaning which helps support Beijing's assertion of its territorial and political claims. Taiwan is part of "China" and the Taiwanese issue is purely a domestic issue of national unification—and explicitly not an international relations concern.

In terms of political and historical supremacy, for the authorities in Beijing, the Republic of China ceased to exist when the CCP declared the establishment of the PRC in 1949. As the following statement mentions:

> There is only one China in the world, and Taiwan is part of China. The Government of the People's Republic of China is the sole legal government of China. This is the fact recognized by the United Nations and over 160 countries, including the United States of America which established diplomatic relations with China 20 years ago.
>
> It was true that in 1912, Dr. Sun Yat-sen led a popular revolution that overthrew the corrupt Qing Dynasty and founded the Republic of China (not the so-called "Republic of China on Taiwan"). What was also true is that in 1949, that Republic of China which had been ruled by a corrupt and despotic Kuomintang regime for 28 years was overthrown in another popular revolution. This is history, whether one likes it or not. Since then, China has been represented in the world by the People's Republic of China with its national government in Beijing, and the "Republic of China" has since become defunct.[20]

It is interesting to note that "including the United States of America" is highlighted in the paragraph, noting that the PRC intended to receive the great power's recognition of her entitlement to "China" and "Chinese government." The phrase "This is history, whether one likes it or not" has also been used in official statements, demonstrating arrogantly that the authorities dismiss all possible queries or criticisms in light of China's historically and politically dominant status over Taiwan. The word "defunct" has been employed to intentionally claim the existence of one single China under PRC rule, despite the fact that ROC had been rejuvenated on the island of Taiwan after the retreat from mainland China in 1949, and was one of the key founding members of the United Nations until 1971.

While ROC is regarded as "vanished" and Taiwan is only a geographical term and even a province of the PRC in mainstream political discourse, a significant number of surfers have shared the identical assertion, as mentioned below:

- "Mainland [China] can be regarded Taiwan as her son, but Taiwan does not appreciate [the mother]. I have something to tell the pro-independents. Independence means war, war leads to the destruction of achievements that have been founded in the past decades." (jackwang1987)
- "What? The Taiwanese never regard themselves as the Chinese? Where does your ancestor come from? If you don't recognize your ancestor (just like you don't recognize your mother), it is inhumane! / If you [the Taiwanese] are competent enough, you can choose not to speak and write

in Chinese. Do you think you write in standard characters [*fanti zhi*] (and you claim that you are non-Chinese)? Standard characters belong to China. Taiwan is not a nation, and where do your national flag? You are not entitled to enter the Olympic Games / I cannot stand [accept] the view as a high school student." (tjlcn)

- "To organize Taiwan's independent movement in the name of Taiwan will be doomed to failure. Love Taiwan and China [*Zhonghua*] are genuine people. The Chinese compatriots cannot be pretended by such running dogs who support Taiwan's independence. The *minnan* compatriots in mainland China and those in Taiwan unite forever! Long live the Chinese people! Guys of the KMT and DPP should go hell! There is only one Taiwan and one China and then the Chinese people unite! There is no space for the pro-independent element in Taiwan." (jodonz)
- "Suggest that [we] do not regard the Taiwanese as the Chinese compatriots. They occupy the island illegally." (su1688888)
- "Take over Taiwan at ease. Using the Gōtō Shinpei's remark: The Taiwanese are greedy, fear death and love face." (fujianyounchun)

Note that the polarization regarding the understanding of relationships between mainland China and Taiwan can be found: On the one hand, a sense of family has been used interpretatively, noting that the ethnical linkage is still in effect despite the political division between the two places. On the other hand, a sense of superiority is imposed, meaning that Taiwan is not so unique since languages and cultures have been borrowed from mainland China. Moreover, China is such an influential country in the region that Taiwan's independence cannot stand in the face of economic and military challenges of the Beijing authorities.

Finally, economic supremacy can be interpreted as Taiwan being basically dependent on mainland China to save, as well as further develop its economy, as reflected in the following remark:

> [Third,] about Taiwan enterprise investment on the mainland resulting in increased unemployment in Taiwan: The fact that the rate of unemployment in Taiwan has increased in the past few years has mainly been caused by a decline of the enterprises' will to investment. In fact, one of the main reasons is that the current leader of Taiwan sticks to the separatist stance of "Taiwan independence," and undermines the cross-Straits relations and impedes the "three direct links," which have dealt a blow to the confidence of Taiwan and overseas investors. At the same time, the rising unemployment rate is also due to its structural unemployment. With the increase of investment and production costs in Taiwan, labor-intensive industries there have further lost their competitive advantages, and investors have had to look elsewhere for low-cost areas for

> continued development. By investing in the mainland, such enterprises can regain their competitiveness, and use the profits they make on the mainland to increase their investment capacity in Taiwan, such as to support the sustained development of the emergent industries in Taiwan, and greatly increase exports to the mainland. All these have played an important role in increasing Taiwan's employment, upgrading its industries and stabilizing its economy.[21]

In the PRC's political discourse, the economic difficulties faced by Taiwan are attributed to domestic chaos, namely the promotion of pro-independent thought and the refusal to further communicate with mainland China through the three direct links. There is only one mainland China that helps Taiwan save itself from economic decline and achieve revitalization. Most important, in the case of any move toward independence, they would not receive any economic benefits from mainland China. Similar opinions can also be found from some surfers, such as:

- "China can take over Hong Kong from Britain, why cannot take over a little Taiwan? Are there any differences of Hong Kong before and after returning to China? Are you afraid of the communist party? Hong Kong continues to enjoy the prosperity after returning to the motherland. People are highly united and have the sense of superiority to China. Apart from the tourist industry, the entertainment industry is closely related to China and Hong Kong. A number of Taiwanese artists have actively developed market in mainland China. Like Chang Hui-mei [N.B. a popular female singer] who performed publicly by singing the national anthem; as a consequence, China and Hong Kong refused her to perform again. It is useless (for her) to befriend with Ah Bian [N.B. Chen Shui-bian]. Taiwan is alike a 14- to 15-year old girl who strives to be independent. After being played [fooled] by Japan and then she turns to the United States." (Chinaheartforever)
- "The current political situation of Taiwan is so chaotic, so the situation would be stabilized by the ruling of mainland authorities. Also, the quick pace of development on mainland, and Taiwan is only a little place in the eyes of mainland China." (jackywang1987)
- "In the 1990s, Taiwan's independents mentioned that [Taiwan] could not unify with mainland China because of poverty. Would they mention this again now? / Needless to say, some coastal provinces on mainland China are richer than Taiwan." (baharlcn)
- "With the increasingly might of China's economy, this promotes Taiwan's economy too. Tourist industry is the key economic activity in Taiwan. Let's think about that. If Taiwan returns to the motherland, the mainland tourists can visit Taiwan without getting a visa. Taiwan can

earn over a billion dollars, apart from mainland's economic investment in Taiwan. China is a money-tree in this regard. . . . With the increasing might of China, the rising status of the Chinese will be valid. With the return of Taiwan to the motherland, the unification of motherland has been completed. These are advantageous to the Taiwanese. After returning, the Taiwanese rule Taiwan." (Chinaheartforever)

- "It is not a long-term plan for Taiwan to isolate herself. Taiwan's return to China is only a matter of time within 5 or 10 years. Taiwan is at present in decline, facing political decay, social polarization, economic decline and the elections that waste money and manpower, turning out to be disappointed. China is now transiting into capitalism, and its economy is taking off. The Taiwanese are also the Chinese compatriots. If Taiwan can be a part of China's economic takeoff, this would be advantageous to Taiwan's economy. Europeans know about the European Union. On one day, you can proud of saying that I am a Chinese and come from Taiwan. . . . If you are strong enough, people would not look you down and you will be respected by others." (Chinaheartforever)

The above quotes share an identical theme: that, ironically, the best way of ensuring Taiwan's economic and political future is for it to become dependent on China. Economically, the Taiwanese economy can be revitalized with the help of mainland China. Politically, a strong and influential economic sphere can be established and developed with the inclusion of Taiwan, and such countries as the United States and Japan would regard China as a great power. In summary, the supremacy of China over Taiwan is to be narrated in the disappearance of the ROC after 1949, the closer historical, ethnic, and cultural linkage with mainland Chinese despite political division, and the territorial status of Taiwan as a province.

Patriarchy

Deploying the familial perspective in viewing Taiwan, some nationalist surfers perceive China as a benevolent senior aiming to demonstrate its fatherhood and/or big brother image to such junior members in power relationships. This can be interpreted by China being willing to help get Taiwan into full swing, provided that Taiwan is expected to demonstrate its submissive mentality to the senior. The following extract shows the brotherhood mentality in interpreting cross-Straits relations, and aims at demonstrating the greatness of China in consummating national unification:

> When Taiwan understands mainland China, it is anticipated that the Chinese in "two shores and three places" [mainland China, Taiwan, and Hong Kong] can

> unite together. It is not a dream that our glorious past in Tang dynasty [A.D. 618–907] can be resumed. That's great. If brothers fight each other, this would humiliate each other. Frankly speaking, we are all the same; it does not a matter of who is the ruler. (fed110)

Second, such a relationship can also be revealed through China vigorously developing in economic and diplomatic dimensions, whereas Taiwan is on the decline economically and diplomatically with the increasing influence of China in regional and international economic systems and in the face of globalization. Actually, a similar pattern can also be seen in the Hong Kong case. Given the decline of Hong Kong after the Asian economic crisis of 1997, Hong Kong seems to have been relying on the help from mainland China to survive. With the increasing dependence on mainland China, some Hong Kong locals address mainland China as grandfather (*Ai Ye*), noting that in case of trouble and difficulty, *Ai Ye* would assist Hong Kong actively, notably through its chief executive, government, or socioeconomic elite, as shown in the introduction to the Individual Visits and the Closer Economic Partnership Agreement (CEPA) after the July 1 rally in 2003. In the face of Tung Chee-hwa's incapable leadership and maladministration, both Beijing and Hong Kong SAR governments believed that economic help would be an effective means to revitalize governmental support. Adopting a similar logic, some nationalist surfers believe that China's assistance to Macau, Hong Kong, and even Taiwan would definitely be a righteous action so as to demonstrate China's might and mercy, as one nationalist surfer has expressed:

> Mainland China can regard Taiwan as her son, but Taiwan does not respect [mainland China]. Hereby I want to address a message to this fool [i.e., Taiwan]. Independence means war, and war would ruin Taiwan's development over the past few decades. Actually, many Taiwanese look down the mainland Chinese, and feel that China is a poor country. Do you know a proverb, saying that "The son would not abandon his mother due to her ugliness. The dog would not look down its master because of his poverty" (jackywang1987).

With due reference to the last sentence, "mother" and "master" have been used interchangeably to denote China, regarding herself as a senior in the face of Taiwan, even though Taiwan does not understand and realize the reality. In fact, the adoption of the familial perspective to view Taiwan has three implications. First, political discourse ensures the maintenance of closer ties between the two sides despite the existing political division. Second, the Taiwan issue is a *de facto* domestic affair given the persistence of the closer political, cultural, and historical connection. Third, the Beijing authorities are justified in taking further action in case the Taipei government takes action to delink the familial tie, notably the declaration of independence. In other words, the mother is reasonable in giving lessons to a naughty son because he behaves poorly, and does not observe family regulations.

In retrospect, hostility, supremacy, and patriarchy are complementary to one another in understanding the nature of Chinese nationalism toward Taiwan. Hostility is a force aimed at drawing the nationalist surfers' attentions to view Taiwan's *de facto* separation from the motherland. They believe that under the presence of the separatist and independent activities on the island, warnings and threats should be imposed and reiterated to restrain the growth and development of this politically undesirable trend. Supremacy and patriarchy are two faces of a coin: a hard dominance is to be imposed on the one hand, in which the separation and independence should be terminated and suppressed, and then cooperation and integration should be encouraged and driven so as to ultimately achieve national unification. By the same token, a soft dominance is concurrently present in shaping the cross-Straits relations, meaning that the Beijing authorities would take care of Taiwan in case of difficulties, with the objective of demonstrating closer ties with the younger member of the family. Taiwan should actively and positively respond to the help from the senior and in return should express a subservient mentality.

Responses

While China has been portrayed as a great power in the eyes of such nationalist surfers, nonnationalists, mainly either pro-independents or pro-status quo surfers in Taiwan, respond to the sharp and harsh criticisms assertively. One example is that while the domestic market is saturated and fiercely competitive, some popular Taiwanese artists, such as Jay Chieh-lun Chou and Jolin I-lin Tsai, also develop a new market in mainland China so as to buttress their popularity in the greater China market. However, some Taiwanese surfers criticize these artists for abandoning Taiwan's identity and shifting into China's identity for the sake of personal business and success. One female Taiwanese band called S.H.E. released a song entitled "Chinese" (*Zhongguo Hua*) in 2007 positively appreciating Chinese culture.[22] However, such a song attracted severe criticism from surfers in Taiwan, saying that S.H.E. "only fawns the Chinese authorities without receiving the expected responses from some of the mainland Chinese surfers. It sings this song for the sake of personal [material] interests instead of racial passion."[23] Also, one of the surfers rewrote the lyrics and retitled the song to "Taiwanese" (*Taiwan Hua*),[24] with the aims of challenging the might of China and reaffirming Taiwanese values. This case reveals that despite the presence of

China's economic attraction, this does not mean that such strength is shared and recognized by some Taiwanese. Instead, the autonomy of Taiwan is weakened in case of depending on it.

Furthermore, even though China is increasingly influential in various facets, it does not mean that China has been fully understood and accepted in the international family, as reflected in the news on the production and exportation of fake goods and food, attracting sharp condemnation from the global media. In this connection, China is a successful, yet irresponsible, businessman in the global market. Such a negative assertion also shapes the ways by which Chinese are being perceived. For example:

- "The quality of the Chinese is low. They like to spit, do not line up while taking the public transportation, do not observe the traffic laws; the traffic light is for reference only in mainland China. When they are walking, they are eating at the same time." (nerostone)
- "When there is a group of people speaking, we know who are the Chinese; and who are the Hong Kongers. The people in the world know that the Chinese like to spit. This is a fact despite the refusal of recognition." (fujiyama123)
- "The Chinese regard themselves are under the rule of the Heavenly dynasty. Most of the Chinese intellectuals are arrogant due to superiority. China is externally strong but internally incapable [*Weiqiang Zhonggan*]." (888 Wayne)

While China emphasizes that Taiwan is part of China, Taiwan is a *de facto* nation in the eyes of some surfers who are likely to be Taiwanese, meaning that nation-building has undergone a unique and severe political transformation from a hard authoritarian regime to a democratic polity. The unification with China could only be achieved by the PRC's annexation of Taiwan by force, resulting in the Taiwanese people once more becoming subject to repressive rule without civic liberties—this time under the PRC.

- "An unequal merging of two countries is annexation. Adopting the military means to recover the sovereignty is invasion. It is a universal value that [we strive to resist] annexation and invasion. The People's Republic of China is a nation with independent sovereignty, Taiwan is not a part of it (even though there is only one minute)." (888wayne)
- "This is a typical way used by Chinese authorities: They label those who are anti-communists as independents; independents are labeled as the American dogs [traitors], and Japanese slaves; and then those who have abandoned their ancestor, so they are sinners." (888wayne)

- "Even though Chen Shui-bian is performed badly, he is elected directly by the Taiwanese. Is there a freedom to vote as well as human rights in mainland China? It is a country which governs in a chaotic manner. Is China a democratic country? Does China observe the rule of law? I accept your opinion that you do not like the Democratic Progressive Party, but you should criticize it in a civilized manner which has demonstrated the poorer attitude." (cat3636)

While it is generally accepted that China is a great power in the regional and international settings, some surfers argue that China is not morally strong, based on their understanding of the social problems associated with mainland Chinese and the psychological handicaps arising from their misconduct.

- "Is China really strong? There are groups of beggars. The polarization between the rich and the poor is terrible. There are some illegal immigrants from mainland China in Taiwan. If Taiwan is backward, why do they still come and look for jobs over there? Also there are groups of brides from mainland China who are being prostitutes. You are really 'strong' and 'progressive' indeed." (MetalBox1300R)
- "Many people, including those who hold an American visa, have a serious psychological problem: Racial inferiority / They embrace the ancestral heritage, such as gunpowder, think that they have been so mighty. However, they faced defeats repeatedly after the Eight-power Allied Expedition (1900), and were almost defeated by the 'little' Japan. After the Second World War, some Chinese are fallen into the trap of racial inferiority and anti-foreignism / They dream of themselves to be 'mighty,' and have achieved something by claiming that 'the Chinese have stood up,' including the detonation of the first atomic bomb, the bidding of the Beijing Olympic Game, and the putting of Chinese astronaut in space. However, the absence of self-confidence has constituted a gesture, noting that a repeated advocacy of 'the standing up of the Chinese' / [Actually,] they need the psychological treatment. When the Chinese think that they are mighty, they would discriminate the Blacks and the Japanese; however, they are afraid of being discriminated. The co-existence of such contradictory mentality can only be healed when they are willing to enjoy an equal status with other countries and ethnicities." (kaiyokenmp)

Overall, such surfers challenge the consistent view held by the nationalist surfers regarding the ultimate unification with mainland China by mentioning the political differences between the two places: An undemocratic and authoritarian mainland China and a democratic and liberal Taiwan cannot be compatible. Also, economic success and achievement cannot make up for the moral decline and decay in Chinese society. Most importantly, such surfers highlight the political identity by positively perceiving the political transformation on the island since the 1980s, despite the incompetent leadership of Chen Shui-bian of the DPP, and the weaker economic performance compared to mainland China and Hong Kong. While the nationalist surfers severely criticize the pro-Japanese/American mentality in Taiwan that aims to alienate Chinese culture, the Taiwanese surfers do not address such criticism. Naturally, the historical influence of both countries in shaping Taiwan's development in the twentieth century cannot be ignored. Additionally, while the nationalist surfers perceive Japan and the United States as imposing their hegemony in Taiwan so that they believe that the Taiwanese are pro-Japanese/American, a number of Taiwanese believe that the popularity of the foreign culture is a hallmark of pluralism and dynamism, rather than an emphasis on de-Chineseness. Above all, the pro-independent or pro-status quo responses regarding online Chinese nationalism toward Taiwan are not so enthusiastic based on the collection of messages, presumably because some surfers think that responses without a meaningful discussion and rational thinking are wasteful.

Conclusion

This study has attempted to understand online Chinese nationalism toward Taiwan by conducting a textual analysis of comments left by surfers, most likely coming from mainland China or mainlanders residing outside China, as well as Taiwanese surfers. In the critical analysis, the mainland Chinese surfers adopt an ethnocentric perspective to understand the Taiwan issue, without considering the Taiwanese perspective. They generally regard the entire national unification as pragmatic and instrumental for the purpose of proliferating a strong and rich China and becoming a great/regional power in the twenty-first century. In approaching the issues relating to Taiwan, this study also highlights that surfers tend to adopt a negative perspective, highlighting Chen's poor leadership and maladministration, and Taiwan's independence and economic decline due to structural transition, but intentionally sweeping aside the democratic transition and consolidation, civil liberties, and rule of law. In responding to divergent views relating to the unification, the logic of the false dilemma is adopted to exclude those opinions questioning the possibility and suitability of unification, and to unconditionally advocate those opinions supporting unification, regardless of the distortion of the social and political reality, or the beautification and exaggeration of the political and economic vision. Of course, the study is not intended to prove the extent to which surfers shape the policy adopted by PRC authorities in dealing with the cross-Straits issue, but to demonstrate that a consensus between state and surfers can exist. However, if surfers' perceptions basically reflect the official stance, online Chinese nationalism would constitute an adverse effect on driving national unification in the sense that such a negative perception of, attitude toward, and response to Taiwan has created mutual antagonism. Such surfers, to a certain extent, are regarded as a group of extremists disseminating arrogance and unfriendly language, and may undermine the political spin undertaken by Beijing authorities to depict China as an open, friendly, peaceful, and civilized nation.

Notes

1. "Asian Internet Usage and Population" [www.Internetworldstats.com/stats3.htm#asia, accessed on 6 November 2009].
2. See Gudran Wacker, "The Internet and Censorship in China," in Christopher R. Hughes and Gudrun Wacker, ed., *China and the Internet: Politics of the Digital Leap Forward* (London and New York: RoutledgeCurzon, 2003).
3. See Christopher Hughes, *Taiwan and Chinese Nationalism: National Identity and Status in International Society* (London and New York: Routledge, 1997); and Jun Toming Liu, "Restless Chinese Nationalist Currents in the 1980s and the 1990s: A Comparative Reading of *River Elegy* and *China Can Say No*," in C. X. George Wei and Xiaoyuan Liu, ed., *China Nationalism in Perspective* (London: Greenwood Press, 2001).
4. Ye Jianying, "Ye Jianying on Taiwan's Return to Motherland and Peaceful Reunification" (1981) [www.china.org.cn/english/7945.htm, accessed on 6 November 2009].
5. Deng Xiaoping, "Deng Xiaoping's Six Conceptions for the Peaceful Reunification" (1983) [lv.china-embassy.org/eng/zt/twwt/t251058.htm, accessed on 6 November 2009].
6. Zhao Quansheng, *Interpreting Chinese Foreign Policy: The Micro-Macro Linkage Approach* (Hong Kong: Oxford University Press, 1986), 127.
7. Susan L. Shirk, *China: Fragile Superpower* (New York: Oxford University Press, 2007), 185.
8. Xu Wu, *Chinese Cyber Nationalism: Evolution, Characteristics and Implications* (Lanham, Md.: Lexington Books, 2006), 155.
9. See Guo Zhongshi, Weng Hin Cheong, and Huailin Chen, "Nationalism as Public Imagination: The Media's Routine Contribution to Latent and Manifest Nationalism in China," *The International Communication Gazette*, Vol. 69, No. 5 (2007), 467–80.
10. Xu Wu, *Chinese Cyber Nationalism: Evolution, Characteristics and Implications*, 123–25.
11. Xu Wu, *Chinese Cyber Nationalism*, 132.

12. Peter Hays Gries, *China's New Nationalism: Pride, Politics and Diplomacy* (Berkeley: University of California Press, 2004), 135.

13. Xu Wu, *Chinese Cyber Nationalism: Evolution, Characteristics and Implications*, 133.

14. Christopher Hughes, *Taiwan and Chinese Nationalism: National Identity and Status in International Society*, 20.

15. Xiwen Gagaga, www.supertv.com.tw/recreation/ ListDate.asp? DramaNo=69.

16. "Zhu Rongji warning of Taiwan's independence" [www.youtube.com/watch?v=8FwjQduSzF8, accessed on 6 November 2009].

17. "Zhu Rongji warning of Taiwan's independence."

18. Elaine Baldwin, Brian Longhurst, Greg Smith, Scott McCracken, and Miles Ogborn, ed., *Introducing Cultural Studies* (Harlow, UK: Pearson Education Limited, 2004), 105.

19. "Taiwan As Part of China Shall Not Change (June 2, 2000)" [english.people.com.cn/english/200006/02/eng20000602_42185.html, accessed on 6 November 2009].

20. "Taiwan—an Inalienable Part of China (December 1998)" [www.fas.org/news/china/1998/981200-prc-tw.htm, accessed on 6 November 2009].

21. "Explanations on Some Questions Related to the 'Three Direct Links' (December 17, 2003)" [www.chinataiwan.org/english/key/tdl/200312/t20031217_128378.htm, accessed on 6 November 2009].

22. "Chinese (*Zhongguo Hua*, April 28, 2007)" [www.youtube.com/watch?v=5IXO3vv4tvE, accessed on 6 November 2009].

23. "Chinese (*Zhongguo Hua*, April 28, 2007)."

24. "Taiwanese surfers: Taiwanese versus SHE 'Chinese' (May 4, 2007)" [www.youtube.com/watch?v=dwFcP1szh4g, accessed on 6 November 2009].

7

The "Two Americas" Dichotomy

*Online Chinese Nationalism Toward the United States**

Yaling Pan

SINCE THE EMERGENCE OF MASS POLITICS in eighteenth-century Europe, nationalism has proven to be a potent force that continues to shape world politics. The rapid development and proliferation of the Internet and other manifestations of information technology has given rise to a novel variation of mass politics: an "electronic democracy" that flourishes in virtual communities.[1] The Internet has not only assisted the Chinese populace in better understanding domestic and international affairs, but has also enabled individuals to exercise independent judgment through participation in "online nationalism." In the midst of its modern-day renaissance, China's online nationalistic fervor has intensified among its 250 million Chinese Internet users (a figure close to one quarter of the entire population). Recent events, including what was widely perceived by the Chinese public as offensive conduct on the part of Carrefour and CNN, incited vehement protests and boycotts. Additionally, the 2008 Beijing Olympic Games concurrently spurred pro-Chinese and anti-Western sentiments.

The most significant dimension of online Chinese nationalism is that expressed vis-à-vis the United States. A rapidly developing nation with great expectations for the future, China is fully cognizant of the United States' enviable status as the world's sole superpower. Chinese Net users are thus naturally drawn to U.S.-China bilateral relations. That being said, from their point of view there are often "two Americas": the "beautiful and advanced America" and the "hegemonic and imperialistic America." The prevailing characterization of the United States at any given moment is contingent upon the state of Sino-U.S. relations at that time.

The "two Americas" dichotomy does not necessarily preclude other attitudes harbored by Chinese Net users with regard to the United States. The majority of Net users are actually ambivalent in their views. Those who profess extreme love or extreme hatred of the United States are decidedly in the minority. Such moderation suggests that Chinese nationalism toward the United States is quite rational, and increasingly so.

The first section addresses the "two Americas" dichotomy and its implications for Sino-U.S. relations. The following section will explore the three types of attitudes outlined above (positive, negative, and ambivalent) while drawing comparisons with the "Two Americas" view. A discussion of the applicability of the three types of attitudes in different realms will follow, as well as an examination of how Chinese attitudes are becoming increasingly rational. Finally, the author concludes that due to the ever-changing nature of Chinese online nationalism, this topic could benefit from future study.

The scope and depth of analysis to be presented is inevitably constrained and complicated by the sheer number of Chinese Net users, the countless number of websites, and the constantly evolving nature of the Internet. Consequently, the research upon which the rest of this chapter is built upon is drawn from the most prominent Chinese online communities, such as the *Qiangguo luntan* (Strengthening Motherland Forum, bbs.cn.yahoo.com), the Sina.com Forum (bbs.sina.com.cn), Yahoo! China Forum (bbsn.cn.yahoo.com), and China's largest virtual community, the Tianya Shequ (Tianya Community, www.tianya.cn). Certainly, there are other well-known and influential virtual communities such as the Tiexue Luntan (The Iron and Blood Forum, bbs.tiexue.net/) and Sina.com's Military Forum (mil.news.sina.com.cn/bbs). However, the views publicized on the former website are noted for their radical nature, while those of the latter are too specialized. The two forums have few active members and are thus excluded from the analysis. In addition to examining the content of various Internet forums, this chapter will also draw on the resources provided by the Horizonkey Group (www.horizonkey.com), a prominent Chinese polling company that occasionally conducts both Internet and phone polls on the thematic topic of Sino-U.S. relations. All resource materials used were collected prior to June 2008.

The "Two Americas" Dichotomy

Chinese perceptions of the United States can first be attributed to their mixed attitudes toward the political culture of the United States. From the Chinese point of view, Americans are fixated with the protection of individual rights and demand a balance of power among the three branches of government.

Americans are cautiously distrustful of the government, and adhere to "a democratic and liberal logic of social life."[2] However, with regard to foreign policy, Americans are purported to only espouse and engage in zero-sum power plays.[3] The Chinese reasoning is as follows. The U.S. domestic experiment in democracy and capitalism may have been highly successful, but its attempts to promulgate its own obdurate conceptions of "democracy and liberty" to the outside world have been decidedly less so. With respect to the American domestic policy, Chinese Net users are inclined to view America through the "beautiful and advanced America" lens. With regard to the American foreign policy though, Net users criticize the United States as a "sinister, imperialistic country" hostile to the PRC.

The "beautiful and advanced America" lens will first be examined. An extensive body of literature presents favorable portrayals of the United States, including a highly influential paper published by Sina Net entitled *Dui Meiguo gai Ai haishi Hen?* (Should We Love or Hate the United States?) The author revels in the fact that after only two hundred years since its birth as a nation, the United States achieved world supremacy as the world's undisputed superpower. In comparison, even after five thousand years of existence, China continues to lag behind the United States on many fronts. The article therefore praises the industrious and innovative spirit of Americans. The American way of life is thus affirmed by the tremendous success the United States has attained.[4]

Another popular text, *Meiguo Zhen Zheng de Kepa Zhi chu Zai Na'er?* (What Is Really So Bad About the United States?) also casts a favorable light on the United States. The author is a former vice commander of the Chinese Air Force, Lieutenant General Liu Yazhou. He writes:

> Although China consists of millions of people who may not love their motherland, these people all love the United States. There are many Chinese officials (from all ranks) who publicly denounce the United States but privately see to it that their own children study overseas in the United States. The attractiveness of the United States is attributed to the following: An elite system characterized by the cadre selection and election systems, the country's largesse and tolerant milieu (the U.S. flag can be printed on one's underwear or be burned on the street), and the profound power of the country's spirit and morality.[5]

Results of several polls have long detected favorable views of the United States as expressed by the Chinese, especially among youths. In 1995, the Chinese Teenager Research Center conducted a poll entitled "Zhongguo Qingnian Kan Shijie (The World in the Eyes of Chinese Young People)." Eighty-seven percent of responders characterized the United States as "wealthy" when asked for their impression of America's material culture, and 89.7 percent of responders believed that America's overall strength was "strong" (table 7.1).[6]

TABLE 7.1
Reasons Behind Chinese Peoples' Deep Impression Toward the United States

Reason	*Percentage**
Economic prowess and developed status	41.6
Wartime and military strength	37.4
Great Power status	24.5
Educated citizenry and commitment to social progress	5.8
Liberty, human rights, and democracy	3.3
Have relatives or friends who have visited America	2.9
Appeal of American products	2.9
Social stability and high standard of living	1.8
Cultural influence of American movies	1.5

*Responders were not limited to one answer. Therefore, the aggregate is greater than 100 percent.
Source: Horizonkey Group, www.horizonkey.com/first/first_jb/594.htm.

A poll conducted by the Horizonkey Group that same year also revealed that 35.3 percent of responders were most deeply impressed with the United States. The United States was considered the strongest power, the richest country, the country that imparts the deepest impression, and the top tourist destination country (three of the top ten of responders' favorite cities were located in the United States).[7]

The Horizonkey Group also conducted a poll entitled "Zhongguo Juming Shenghuo Zhishu Diaocha (Chinese Residents' Life Index Investigation)" in April 2001 and October 2002. The two polls together revealed that 10 percent of Chinese residents believed that the United States was the friendliest country toward China (the United States thus rose to No. 6).[8] In another poll held in June 2002, when asked "Which country are you most deeply impressed with?" 26.5 percent of responders selected the United States. In a follow-up question inquiring the reason, 41.6 percent attributed their deep impression toward the United States to the country's affluence and status as an advanced developed nation. To the question "Which country comes to mind when the words 'wealthy' and 'strong' are used?" 41.6 percent and 67.0 percent of responders respectively chose the United States in 1995. In 2003, the figures rose to 76.2 percent and 82.6 percent, respectively.[9]

To summarize, many Chinese regard the United States as a "beautiful and advanced country" due to the United States' admirable domestic policies, political system, social organization, and inherent values.[10] There are several additional reasons.[11] For example, the American education system, especially at the tertiary level, is the envy of the world.[12] The American emphasis on the individual is also greatly admired. Many Chinese believe that China has much to learn from emulating the United States.

The "hegemonic and imperialist" lens will now be examined. The paper *Dui Meiguo gai Ai haishi heng?* (Should We Love or Hate the United States?) not only provides a favorable depiction of the United States (as discussed earlier), but a negative portrayal as well. The article denounces the United States' belligerent posturing vis-à-vis other countries, including its excessive meddling in China's efforts at unification. In an attempt to understand the U.S. opposition to Chinese unification efforts, the United States' unpalatable experience during its own civil war (1860s) is often brought up. Many also dismiss Washington's fears of China's military given that the American defense budget often amounts to more than the total sum of other countries combined. Ultimately, the author of the paper posits that the desire to conquer lies deep at the heart of American culture. Therefore, while harmony and welfare are stressed domestically, the U.S. foreign policy reflects an intrinsic drive to plunder, kill, and conquer abroad. Scores of Chinese dismiss America's "universal appeal," and instead lambast the egocentricity of such thinking.[13] The Chinese view of a "hegemonic and imperialist America" also stems from the U.S. stances regarding the Taiwan issue, human rights, arms sales, and trade. In short, many are disconcerted by the U.S. application of hypocritical double standards and its myopic reluctance to acknowledge competing but equally valid international principles and behaviors.[14]

The U.S.-waged Iraq War of 2003 further galvanized Chinese opposition to American international hegemony. The United States had justified the war based on the existence of weapons of mass destruction, but Chinese Net users remained unconvinced of the rhetoric. Chinese websites widely speculated on the true causes behind the American invasion. The American desire to secure oil reserves, the aspiration to dominate the Middle East, and the goal of ultimately conquering China via an invasion of North Korea were all suggestions that circulated on the Web.

The 2003 Iraq War was widely opposed by the Chinese populace, who believed that the war constituted a severe breach of state sovereignty. According to a 2003 poll conducted by Horizonkey Group, 79.7 percent of responders were against the Iraq war, 8.2 percent supported it, while 12.1 percent were indifferent.[15]

The results of another Horizonkey poll that elicited the participation of 445 residents of Beijing, Shanghai, and Guangzhou conveyed the Chinese people's strong opposition to the war. 49.2 percent of responders voiced "strong disapproval" of the U.S. conduct while 41.3 percent expressed "disapproval." Only 9.5 percent expressed "approval" of the U.S. behavior; 2.6 percent articulated their "strong approval." Within a five-score measure, America's appreciation score dropped precipitously from a score of 3.16 in 2002 to a score of only 1.73 in 2003. In comparison, more than 60 percent of responders approved

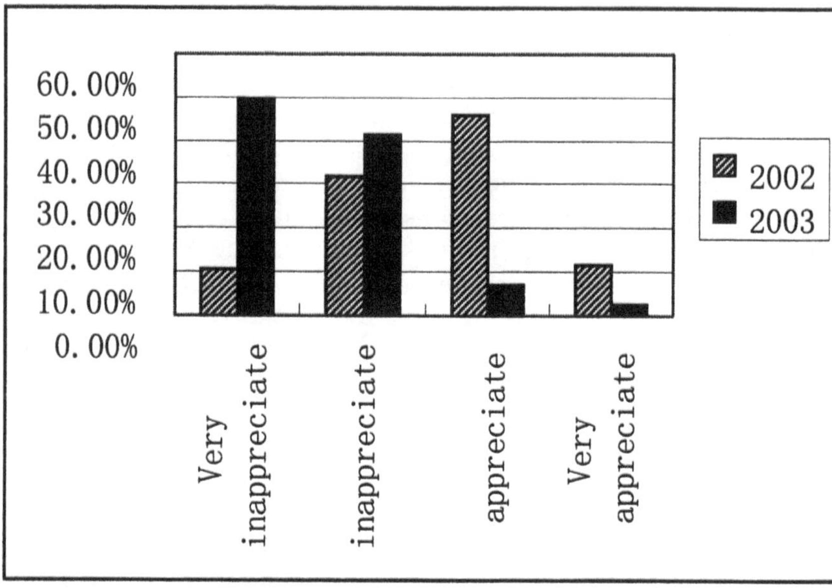

FIGURE 7.1
Chinese Residents' Views on the United States: 2002 and 2003
Source: "Fangan Zhanzheng: Zhongguo ren mei de shuo! (Anti-War: Chinese, absolutely!)," Horizonkey Group, Nov. 14, 2003, available at www.horizonkey.com/showart.asp?art_id=241&cat_id=6.

of Iraq's conduct (11.6 percent strong approval; 47.3 percent approval). Only 40 percent of responders disapproved of Iraq's conduct (30.2 percent disapproval; 9.1 percent strong disapproval). Iraq's score rose from 2.41 in 2002 to 3.20 in 2003 (figure 7.2).[16] Chinese approval of Iraq's conduct most likely arose from sympathy for a country whose territorial sovereignty was perceived to be egregiously violated, whereas Chinese disapproval of the United States resulted from the country's decision to initiate war.

In the same poll, more than 90 percent of responders used words with negative connotations to describe their perception of the United States. The most frequently employed word was "hegemony" (51.9 percent). Other words, phrases, and epithets employed included "unreasonable," "bad," "bully of the weak," "extremely conceited," "brutal," "violator of human rights," "aggressor," "violent," "belligerent," and "egotistical." Some even called the United States a "bandit" and a "gangster." In comparison, the few words, phrases, and epithets with positive connotations included "advanced," "good," "appreciable," "vast territory and abundant resources," "inclusive," and "generous."[17]

A fundamental factor that invigorates the "hegemonic and imperialist" view of the United States lies in the basic fact that many Chinese do not appreciate the United States' implicit and (often) discreet remarks and criticisms

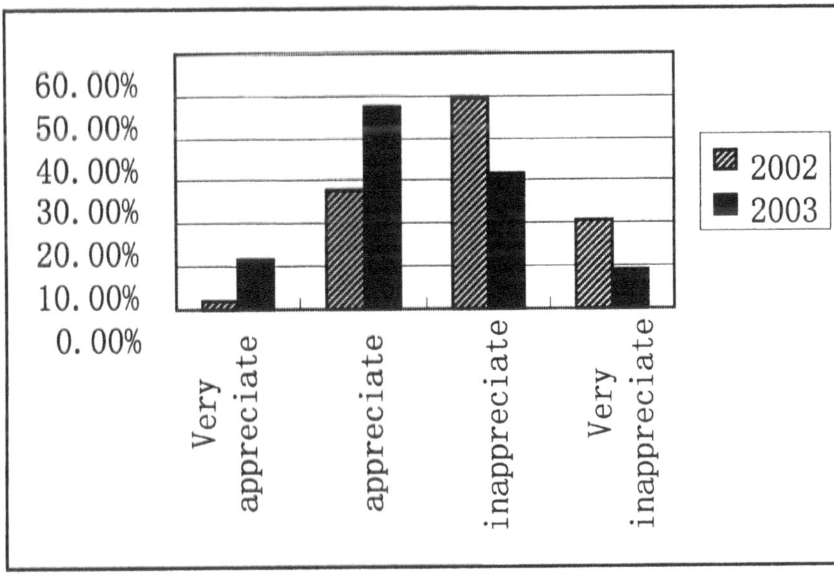

FIGURE 7.2
Chinese Residents' Views on Iraq: 2002 and 2003
Source: "Fangan Zhanzheng: Zhongguo ren mei de shuo! (Anti-War: Chinese, absolutely!)," Horizonkey Group, Nov. 14, 2003, available at www.horizonkey.com/showart.asp?art_id=241&cat_id=6.

directed toward China and other countries. The results of numerous polls indicate that many Chinese believe that the United States exercises economic, political, and cultural hegemony and is a country that coerces other nations to accept American values and institutions.

In a poll conducted by Horizonkey Group in 2005 entitled *The World in Chinese Eyes*, 85 percent of responders believe that the United States "intervenes in other countries' domestic affairs," 83 percent believe that the United States "tries to economically dominate other countries," and 68.8 percent state that they associate the United States with hegemony (figure 7.3). With regard to America's involvement in East Asia, 46 percent of respondents believe that such engagement is inimical to Chinese national interest, a statistic significantly higher than the positive response (36.1 percent) (see figure 7.4). 55.6 percent of responders believe that the United States disrespects China, 44.5 percent believe that the United States exhibits "some disrespect," while 11.1 percent think that the United States exhibits "great disrespect" (figure 7.5).

Of course, Chinese Net users' perception of the United States is ever-changing and is often dependent on the economic, political, and diplomatic state of relations between the two countries. When bilateral relations sour, Chinese Net users' express their great disdain and even resort to drastic

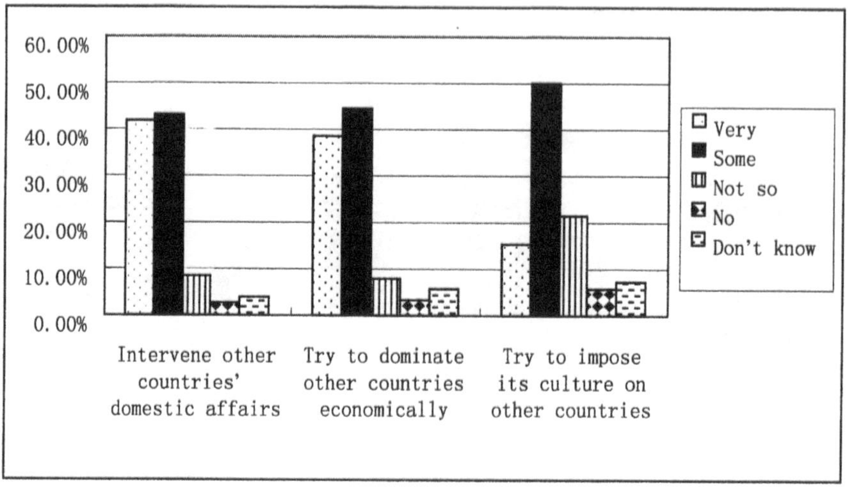

FIGURE 7.3
Chinese Residents' Perceptions of the United States

Source: Horizonkey Group, *The World in Chinese Eyes Poll, 2005*, Nov. 17, 2005, available at www.horizonkey.com/showart.asp?art_id=480&cat_id=6.

measures such as Internet hacking. In contrast, during periods of positive bilateral relations, Chinese Net users openly praise the U.S. democratic and education systems.

In the 2006 poll of *The World in Chinese Eyes*, Horizonkey interviewed 3,167 individuals residing in ten of China's largest cities (Beijing, Shanghai, Guangzhou, Wuhan, Chengdu, Shenyang, Xi'an, Dalian, Xiamen, and Ji'nan). Also, 611 public opinion leaders comprising experts, businesspeople, media

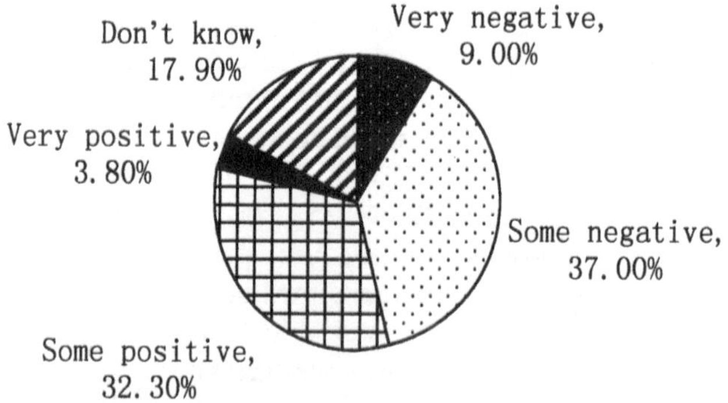

FIGURE 7.4
Chinese Residents' Responses to "Does American Engagement in East Asian Affairs Exert a Positive or Negative Impact on China's National Interests?"

Source: Horizonkey Group, *The World in Chinese Eyes Poll, 2005*, Nov. 17, 2005, available at www.horizonkey.com/showart.asp?art_id=480&cat_id=6.

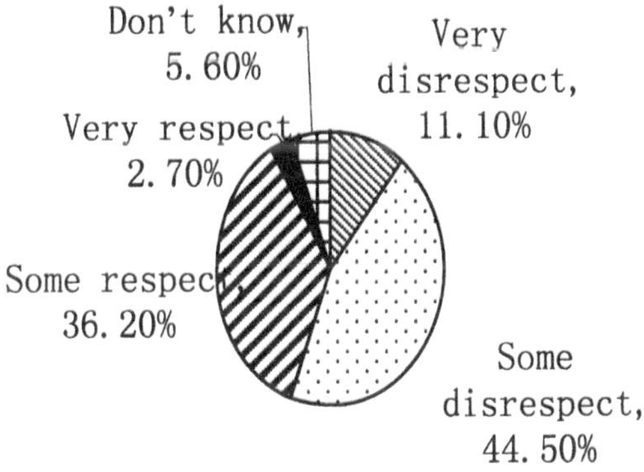

FIGURE 7.5
Chinese Perception of the United States' Level of Respect Toward China
Source: Horizonkey Group, *The World in Chinese Eyes Poll, 2005*, Nov. 17, 2005, available at www.horizonkey.com/showart.asp?art_id=480&cat_id=6.

reporters, and undergraduate university students were asked to provide their opinions. The results of the poll suggest a great amelioration in Sino-American relations that surpassed the state of relations attained in 2001 (thought to have been the pinnacle of bilateral relations). That being said, Chinese public opinion leaders declared that both cooperation and conflicts would characterize bilateral trends in the near future.[18]

Under George W. Bush's administration, Chinese public opinion toward the United States plummeted greatly during the first three years, but gradually improved afterward. A consequence of economic, political, and diplomatic tensions, bilateral relations reached their nadir in 2003. After 2003, however, strategic dialogues and high-level visits assuaged bilateral frictions. Chinese public opinion toward the United States improved from the historical low of 2.36 in 2001 to a score of 2.57 in 2006 (at 2.79, the score of public opinion leaders was nonetheless still higher). 66.78 percent of public opinion leaders articulated a "highly favorable," or "favorable" view of the United States, a figure that is significantly higher than that of the general public (figure 7.6).

Favorable Chinese public attitudes toward the United States as indicated by the 2005 poll of *The World in Chinese Eyes* reflect developments that occurred in 2004, one of the best years in terms of bilateral relations. That being said, prevailing attitudes still remain somewhat negative due to lingering concerns about the U.S. criticism of China and its national policies.[19]

A five-point scale was used to gauge the Chinese public's perspective of the United States, with a score of 1 indicating unimportance/lowest approval, and a score of 5 indicating importance/highest approval. Bilateral relations were given a score of 3.41 in 2005, slightly lower than the highest score attained (3.51) in the fall of 2002. 66.3 percent of respondents therefore

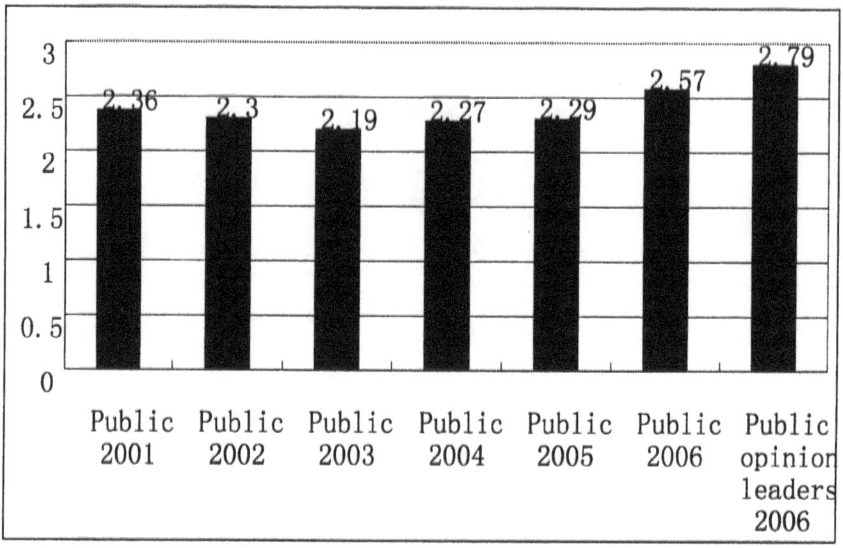

FIGURE 7.6
Chinese Public's Favorable Attitudes Toward the United States (2001–2006) and Public Opinion Leaders' Attitudes Toward the United States (2006)

Source: "Zhongguo Gongzhong dui Mei Haogan da fu Tisheng (Chinese Public Much Favor the United States)," Horizonkey Group, Dec. 14, 2006, available at www.horizonkey.com/showart.asp?art_id=609&cat_id=6.

evaluated Sino-American relations in a positive light. Additionally, the majority of the Chinese public (91.6 percent) believes that Sino-American relations are important, a sentiment reflected in the score of 4.09. 69.2 percent of respondents believe that the United States is the most important country regarding its influence on Chinese economic development. Nonetheless, Chinese public attitudes toward the United States were lukewarm at best in 2005 (score of 2.86) when 54.5 percent of respondents pronounced moderate and strong disapproval of the United States. This was marginally better than the scores received in 2003 (2.83) and 2004 (2.84). See figure 7.7.

Online Chinese Nationalism toward the United States: Three Types of Attitudes

Given the rapid economic development and increasing influence of China, Sino-American relations have become more and more complex. Not surprisingly, most Chinese Net users have rather complicated perceptions of the United States. Many express an ambivalent love/hatred toward the United

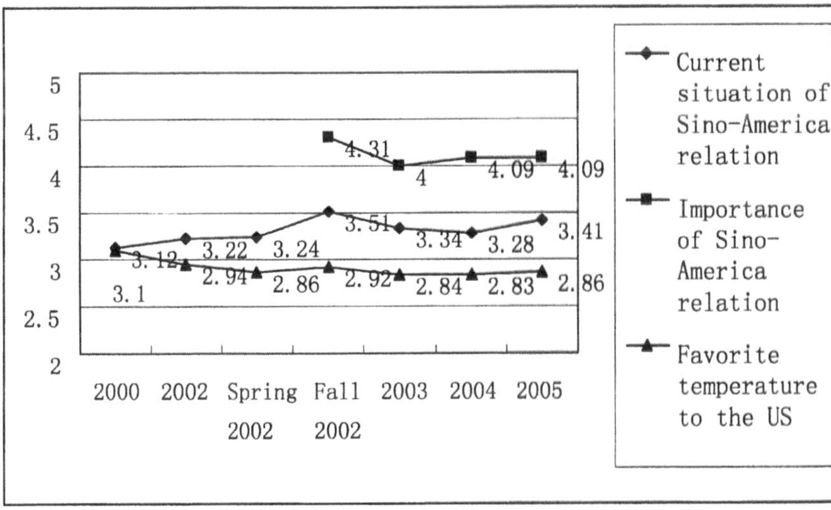

FIGURE 7.7
Chinese Public's Evaluation of Sino-American Relations

Source: "Zhong Mei Guanxi jian ru Jiajin, dan Mingzhong Haogan du ren jiao di (Sino-American Relations Better Up, but Favorite Temperature Still Low)," Horizonkey Group, June 2005, available at www.horizonkey.com/showart.asp?art_id=480&cat_id=6.

States, and lean to one extreme or the other depending on whether the U.S. domestic or foreign policy is the subject of discussion.

Great disdain (hatred) of the United States will first be analyzed. The preponderance of Chinese hatred toward the United States is directed toward the latter's foreign policy. Table 7.2 presents various reasons that are responsible for this phenomenon. Chinese perceptions can be rather fickle, as demonstrated by Chinese attitudes toward the United States immediately before and after the United States' accidental decimation of the Chinese embassy

TABLE 7.2
Key Reasons Behind Anti-American Sentiments

Reason	*Percentage*
U.S. criticism of human rights in China	26.9
U.S. foreign policy	26.1
U.S. role concerning the Taiwan issue	25.2
U.S. policies concerning Chinese accession to the WTO	10.9
U.S. criticism of China's family planning policies	3.4
Other	6.7
Don't know	0.8

Source: "Meiguo ren ruhe kan Zhongguo (How Americans View China)," Horizonkey Group, Feb. 14, 2001, available at www.horizonkey.com.

in Yugoslavia on May 8, 1999. Prior to the incident, a poll conducted by Horizonkey Group revealed that 55.6 percent of respondents held a "favorable" view of the United States while only 9.6 percent held an "unfavorable view." However, in the immediate aftermath of the bombing, only 1 percent of the responders still held a "favorable" view of the United States, while the percentage of respondents professing an "unfavorable" view skyrocketed to an unprecedented 93.8 percent.[20]

The table above presents several key reasons behind Chinese anti-Americanism. It is clear that the U.S. role with regard to Taiwan-Chinese relations remains a highly sensitive issue that fosters anti-Americanism. Due to space constraints, the Taiwan issue cannot receive the in-depth analysis it deserves. Other incidents will be presented and discussed instead.

In late 2007, the American *Kitty Hawk* aircraft carrier failed to visit Hong Kong, precipitating an abrupt dip in Sino-American relations. Some Chinese Net users argued that the failure of the ship to visit Hong Kong was intentional. The most prolific text quoted in Sina Forum in the wake of the incident was an article entitled *Xiaoying Hao Shijian Jiekai Zhongguo Junshi Fangong de Xumu!* (The *Kitty Hawk* Event Unveils the Prologue of the Chinese Military Counter-attack!) The argument presented in the article is as follows:

> After misleading the entire world with respect to the Iraqi and Iranian debacles, the United States now considers China its "imaginary enemy." The United States wishes to "Xiuli China" (to give China a lesson), suggesting political and/or military invasions. China should keep abreast of all developments and be ready to deflect American advances. The United States always needs an "enemy" in order to sustain its military-industrial complex. America has a long history of trying to teach other countries a lesson, as demonstrated by its (prior) engagements in Vietnam, Iraq, and Iran. It is only after one comprehends the "Xiuli (*country name*)" rationale that one realizes that the recent *Kitty Hawk* debacle was no accident, but a result of the Chinese government's inability to tolerate the United States' belligerence any longer. As a peaceful, harmonious society, China must be willing to stand up to the United States. Once China is regarded as an "imaginary enemy," the United States will never stop trying to "Xiuli China" nor will it cease in its efforts to intervene in China's domestic affairs politically and even perhaps militarily. The *Kitty Hawk* incident is just the beginning. China must be aware of the United States' advances. It's time to say "no" [to the United States] and disclose the United States' grand strategy of conspiracy for the entire world to see![21]

Another notable example of an event that ignited a deluge of negative Chinese sentiments toward the United States occurred during CNN's coverage of the Tibetan turmoil that took place on March 14, 2008. In a report by

CNN host Wolf Blitzer that compared contemporary China to the China of the past, CNN commentator Jack Cafferty outraged Chinese viewers when he provided the following response to a question:

> I don't know if China is any different, but our relationship with China is certainly different. We're in hock to the Chinese up to our eyeballs because of the war in Iraq, for one thing. They're holding hundreds of billions of dollars worth of our paper. . . . We are also running hundreds of billions of dollars' worth of trade deficit with them, as we continue to import their junk with lead paint on them and the poisoned pet food and export, you know, jobs to places where you can pay workers a dollar a month to turn out the stuff that we're buying from Wal-Mart. . . . So I think our relationship with China has certainly changed. . . . I think they're basically the same bunch of goons and thugs they've been for the last fifty years.[22]

The response above, particularly Cafferty's characterization of Chinese people as "goons and thugs" greatly offended not only the Chinese government, but the Chinese people at large. CNN's initial apology claimed that the affront was directed exclusively toward the Chinese government, not the Chinese people.[23] However, many Chinese people believed otherwise, and denounced CNN's botched attempt to justify Cafferty's remarks.

In the wake of the comments, several Net users created a website dubbed "Anti-CNN." Soon after its launch, the website was blocked by the U.S. government. The Sina Forum organized a signature petition protesting CNN, and entitled the petition "Questioning CNN's Qualifications for Reporting the 2008 Beijing Olympic Games." Among the more than 890,000 signatories, thirty thousand were overseas Chinese. Among the overseas Chinese signatories, more than six thousand resided in the United States.[24] The number of signatories to the petition grew steadily until the Sichuan earthquake of May 12, 2008.

Furthermore, a paper entitled "Chong CNN kan Meiguo dui Yazhou ge Guo Taidu (America's Attitude toward Asia from CNN's Perspective)" in Tianya Community (the largest Internet community in China) argues that the symbols CNN supposedly uses to represent Asian countries are inherently unfair. The following is illustrative of the paper's central arguments:

> North Korea is represented by a leather shoe. Does this mean America will trample North Korea with an iron shoe? Pakistan is represented by a fist. Does this mean America will deal with it using an iron fist? China is represented by an open door. Does America view China as a country into which it can simply enter, and consider it a prize to be won?[25]

Another Net user uploaded a fake article, entitled "Mei Xuanchuanbu Yanshu Chuli CNN Zhuchiren de Jueding (Decision to Punish Host of CNN Seriously by Department of Propaganda of the United States)." It stated:

> After a comprehensive investigation, the Department has determined that Cafferty's comments were irresponsible. His remarks came at the wrong time and place, violating the *Guanyu zai Aoyun qijian bu de Jingxing Fu mian Baodao de Guiding* (Rules prohibiting Negative Reporting during the period of the Olympic Games) as dictated by the Department of Propoganda of the United States. The comments were highly inimical to the Sino-American friendship and damaged the international image of the American government. As a result, the Department has decided to punish Cafferty and relevant persons involved.[26]

The paper was obviously a fake. It possibly (intentionally or unintentionally) satirizes the Chinese government itself. That being said, it nonetheless reflects Chinese Net users' clear indignation toward CNN.

The harshest condemnation of CNN was summarized in a widely circulated sentence: "Heku wuakong xinshi, pian yao longjiachengzheng, zuoren bu neng tai CNN (Why rack your brains, why must you be specious, why can't you conduct yourself properly, CNN)?"

Lying on the other extreme of the spectrum are those who love the United States to the point that they desire to live there. From their perspective, the United States is a powerful, advanced nation whose material life is superior to that of China. Additionally, these individuals profess a profound admiration for American democracy and the equal opportunities afforded to all.

The number of Chinese immigrants to the United States has been increasing in recent years. A key reason is that Chinese immigrants profoundly admire the entrepreneurial and competitive spirit of the United States. In a 2004 poll conducted by Horizonkey that elicited the participation of 1252 residents of Beijing, Shanghai, Guangzhou, Wuhan, Chengdu, Shenyang, and Xi'an, 20.2 percent of respondents expressed the greatest desire to live in the United States among five given countries. This figure is only slightly behind the percentage of those who chose China (21 percent). The breakdown of the most significant factors at play is as follows: national economic competitiveness (28.9 percent), environmental considerations (25.0 percent), social culture (14.9 percent), customs and traditions (7.9 percent), and national identity (7.8 percent). The United States outperformed its competitors in the area of economic competitiveness (68.3 percent) and social culture (23.5 percent).[27]

The American educational system is another attractive feature of the American way of life. Many Chinese have conjectured on the factors propelling America's continued success in the education sector. One reason is that

TABLE 7.3
Key Factors that Attract Chinese to Live Abroad

China	*United States*	*Australia*	*Canada*	*France*
Social Culture (37.7)	Economy (68.3)	Environment (67.2)	Environment (63.2)	Social culture (58.4)
Nationality and Identity (36.9)	Social culture (23.5)	Economy (17.0)	Economy (24.0)	Environment (26.0)
Optimism towards the Future (6.9)	Environment (5.9)	Social Culture (10.3)	Social Culture (12.7)	Always Favorite (7.9)
Kinship (5.5)	Other (0.9)	Other (5.5)	Other (2.1)	Economy (4.4)
Other (5.1)	Don't know (1.4)		Don't know (1.6)	Other (2.4)
Don't know (7.9)				

Source: "Meiguo Juzhu Jinzhengli Qiangjing, Zhongguo ren ai hen liang nan quan (American Inhabitation Competitiveness Strong, Chinese Love and Hatred)," Horizenkey Group, Oct. 11, 2004, available at www.horizonkey.com/showart.asp?art_id=377&cat_id=6.

the education is financed by taxpayers, and that every child is entitled to a free primary and secondary education. The U.S. education system not only instills civic virtues in its students, but enables each child to reach his or her maximum potential. Additionally, those who want an education (regardless of age, race, religion, or geographic region) are able to enroll in suitable programs with relative ease. Furthermore, the American education system at the university level is the envy of the world. Finally, the American academic community has long established high standards that emphasize the pursuit of free inquiry and academic integrity.[28]

Some Chinese also believe that the Chinese style of education is too rigid compared with the flexibility of the American system.[29] The following is a passage that contrasts the American elementary school with its Chinese counterpart:

> American elementary schools do not simply expect children to regurgitate information in the classroom, but rather emphasize the need [for children] to explore outside of the box. Teachers do not ask students to recite formulas and theorems, but instead teach students how to think and provide them the skills necessary for individual problem-solving. They do not classify their students along academic lines of best, mediocre, and poor performance. Instead, teachers praise students for their efforts and try to inspire them.[30]

The third and most common stance of online Chinese nationalism is ambivalence toward the United States. A poll of undergraduates revealed that the majority were ambivalent toward the United States, admiring the "beautiful and advanced" aspect of the United States while condemning the "unilateralist and hegemonic" aspect. A Chinese software engineering student is quoted as saying that when China ultimately embraces the United States' rigorous anti-piracy policies, opportunities in his field will open up. Until then, China's software industry remains comparatively undeveloped. Another undergraduate student confesses that although he expresses anti-American sentiments publicly, he secretly plans to continue working hard so that he can one day study in the United States.[31]

A demographic group whose members often espouse an ambivalent attitude toward the United States is the returnees, or the *Haigui*. Although their extensive sojourns abroad afforded them the opportunity to fully experience American life, many still profess their deep affinity toward their nation of birth. A poll conducted in 2003 indicates that the number of returnees is increasing at a rate of 13 percent every year.[32] Due to their unique firsthand perspective of American life, the *Haigui*'s perception of America exerts a significant influence on the Chinese perceptions of America at large.

Many Net users appear to rather enjoy criticizing the United States. According to a Horizonkey poll, there are three defining categories that shape Chinese Net users' perceptions of the United States. First, many Chinese regard the United States as a wealthy and strong power troubled by war and conflicts. Second, many Chinese are pragmatic in their approach with China and only absorb the most advanced and useful components of American life. Third, although the United States remains the top foreign destination of the Chinese people, China's rapid ascension and growing strength inevitably attenuates the attractiveness of the United States.[33]

Net users have also embraced a rather cynical view of the United States. For example, NBA star Yao Ming is a household name in China. Yao may be über-famous, but cynics are reluctant to celebrate his success. In a piece entitled "Meiguo ren yong Yao Ming huan zhou le Shenme? (What have the Americans Gained from Accepting Yao Ming?)," the author posits that Yao's entry into the NBA has elicited an unprecedented interest in the NBA among the Chinese populace. In fact, the NBA is often more popular than homegrown sports teams. Cynics thus point out that Yao's recruitment to the NBA merely secured for American business interests a large market share in China. China only receives Yao Ming's income tax, but the United States enjoys significant profits for having secured a loyal and ever-growing fan base in China. Essentially, they argue that the United States disproportionately benefits from Yao Ming's participation in the NBA.[34]

Additionally, although most Chinese praise the American education system, there are those who question its merits. In a piece entitled "Meiguo Xuesheng neng gan Shenme jiu gan Shenme ma? (Can American Students Simply Do What They Want to Do?)," the author attacks prevailing beliefs about the American education system. The author posits:

> The deepest misunderstanding regarding the American education system is that students are allowed to do whatever they want. Let's think this through carefully. America may be a free society, but is still a society governed by the rule of law. American children do not become civil and disciplined citizens if they are not subject to the strict guidelines set forth at home and at school. The primary emphasis of American schools is to instill good citizenship in their students, and therefore students are subject to discipline. All American schools have defined rules of behavior and punishment. Those guilty of infractions are labeled "defiant," "rebellious," "disobedient," "disrespectful," "aggressive," etc. Therefore, American students simply cannot just do whatever they want.[35]

Many Chinese are pragmatic in their approach to the United States, absorbing only what they deem useful and modern. In a Horizonkey poll conducted in 2003, when asked which brand they considered the most famous of American brands, respondents chose Coca-Cola (18.4 percent), McDonald's (18.4 percent), Microsoft (12.1 percent), Kentucky Fried Chicken (10.3 percent), Nike (10.2 percent), GE (10.2 percent), IBM (7.0 percent), Ford (6.4 percent), Lincoln (3.1 percent), and Buick (3.0 percent). It is evident that due to their prevalence and relevance to quotidian life, most Chinese are most familiar with the above-mentioned trademarks of the American clothing, car, food, and high technology industries. Given the current era of globalization, many Chinese are cognizant of the need to pragmatically interact with and learn from foreign cultures.[36]

Toward Rationality

The Chinese translation of the United States is *Meiguo*, which literally means "beautiful country." Despite the seemingly benign connotation of this name, Sino-American relations are actually quite complex. That being said, given the extensive dialogues between the two countries, the Chinese view of the United States can be considered to be increasingly rational. In the period in which the Chinese Yugoslavia embassy and the EP-3 incidents transpired, the Horizonkey Group conducted three polls in October 2001, April 2002, and October 2002. Despite their disapproval of American military practices, the Chinese people in all three polls believed that the United States remained China's most important country with which to interact with. Furthermore, they rejected the notion that globalization is exclusively synonymous with Americanization. 36.7 percent of respondents also believed that Sino-American relations were moving in the right direction,[37] and 82.7 percent of respondents believed that the United States remains China's most important trade partner.[38] The Chinese people therefore tend to differentiate between political and economic developments to arrive at different subconclusions about Sino-American relations.

Chinese aversion to American foreign policy does not extend to the realm of consumerism. The Chinese vociferously consume American products, watching Hollywood movies and eating McDonald's French fries.

Rational views of the United States from the Chinese perspective have become manifest in several ways. First, an increasing number of Chinese (mainly comprised of enlightened undergraduate students) have recognized that realist *realpolitik* remains a major force that propels international relations. Both the United States and China pursue their own respective self-interest. Therefore, it is only natural that when such interests are incompatible, the more powerful nation (the United States) most often proves to be triumphant.[39]

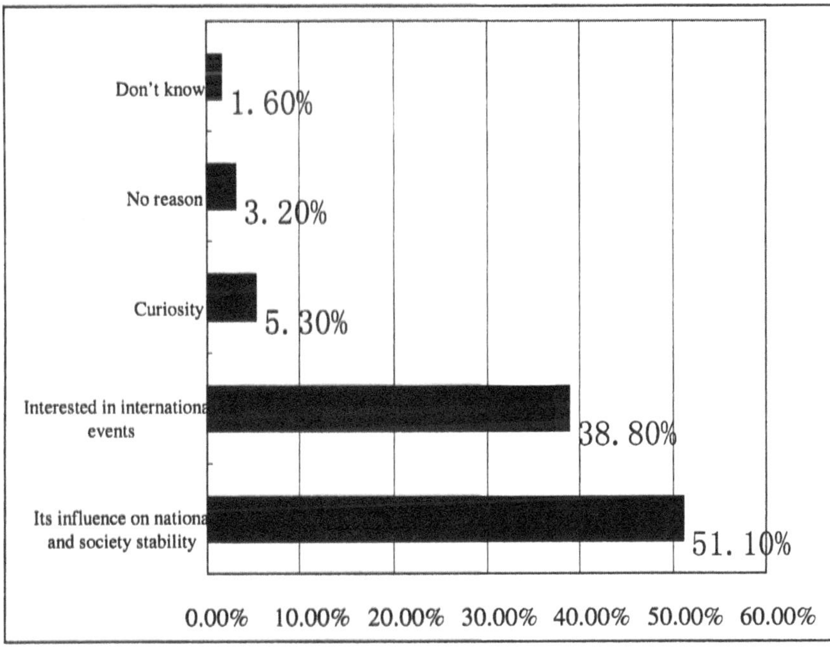

FIGURE 7.8
Chinese Net Users' Interest Level in Terrorism
Source: "Wangming Pupian Guanzhu Kongbu Huodong, Quanming Lianhe Fankong zhui shou Zhancheng (Net Users Pay Attention to Terrorist Activities, Favor Cooperative Anti-Terrorism)," Horizonkey Group, Nov. 30, 2002, available at www.horizonkey.com/showart.asp?art_id=216&cat_id=6.

Following the events of September 11, 2001, anti-terrorism efforts became a top priority of America's global strategy. In order to explore Chinese Net users' views on terrorism and anti-terrorism tactics, the Horizonkey Group conducted a poll held November 22–27, 2002, in cooperation with Sohu web. Ninety-five percent of the 6008 respondents reported that in light of the 9/11 attacks on the United States, they now pay closer attention to terrorism-related developments. Additionally, more than 50 percent of respondents reported that they also accord greater interest toward terrorism's deleterious ramifications on national and societal stability.[40]

With regard to the most effective anti-terrorism tactic, Net users most favored "united anti-terrorism efforts through the mobilization of all people," "military strikes," and "international cooperation."[41]

China's rise has elevated the confidence of the Chinese people, but has also exerted a rational influence on their perceptions of the United States. Recent developments that have greatly boosted Chinese nationalism include the return of Hong Kong and Macao, China's accession to the World Trade

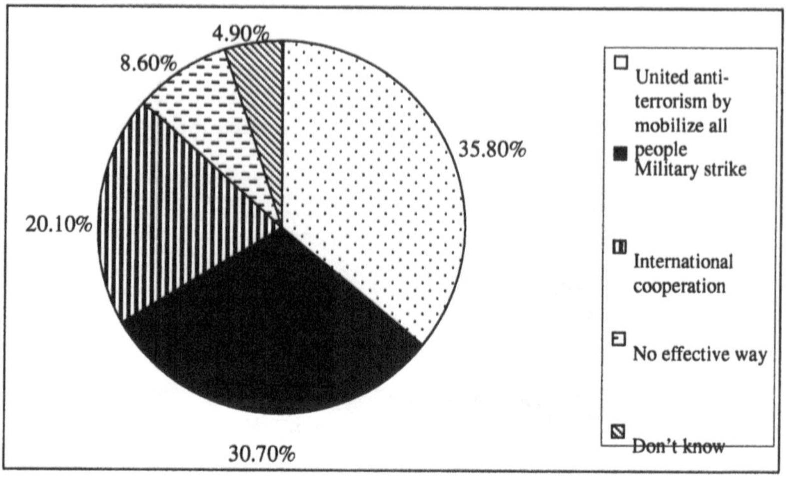

FIGURE 7.9
Chinese Perceptions of the Most Effective Anti-Terrorism Measures
Source: "Wangming Pupian Guanzhu Kongbu Huodong, Quanming Lianhe Fankong zhui shou Zhancheng (Net Users Pay Attention to Terrorist Activities, Favor Cooperative Anti-Terrorism)," Horizonkey Group, Nov. 30, 2002, available at www.horizonkey.com/showart.asp?art_id=216&cat_id=6.

Organization, the 2008 Beijing Olympic Games, and the yet-to-be held 2010 World Expo in Shanghai. A comparison of responses to a Horizonkey poll reveals that the Chinese peoples' impression of their own country has improved drastically from 1995 to 2003. In 1995, only 13 percent of respondents reported the deepest impression toward China, a figure which climbed to 37 percent in 2003 (much higher than that of the United States, 26.5 percent, and France, 5.1 percent).[42]

TABLE 7.4
Countries Exerting the Deepest Impression on Chinese Public (1995 and 2003)

1995	*2003*
United States (35.3%)	China (37.4%)
Japan (25.6%)	United States (26.5%)
China (13.1%)	France (5.1%)
Singapore (3.7%)	Japan (3.8%)
Germany (2.7%)	Australia (3.2%)
Former Soviet Union (2%)	Iraq (3.1%)
France (1.8%)	Britain (1.8%)
Canada (1.7%)	Germany (1.7%)
Switzerland (1.6%)	Canada (1.5%)
Australia (1.4%)	Singapore (1.4%)

Source: Sheng Yu, "Guojia Yingxiang: Zhongguo ren zhui ren Zhongguo (National Image: Most Chinese Identify with China)," Horizonkey Group, Nov. 14, 2003, available at www.horizonkey.com/first/first_jb/593.htm.

Although a minority of Net users were infuriated with foreign countries' actions prior to the 2008 Beijing Olympic Games, the overwhelming majority of Internet communities called for restraint and urged Net users to exercise rational patriotism. Such suggestions included

(1) It is good to love your country, but harboring too much passion will harm your country!
(2) The people welcome all manifestations of patriotism, yet are strongly against behaviors deleterious to national security and those that underscore national prestige.
(3) To love your country, to oppose foreign interventions, and to pursue national reunification are all rational feelings.
(4) The best way to demonstrate your patriotism is to protect the national interest and to eschew damaging national security and prestige.
(5) Since we are loyal [citizens], we will boycott quietly and will not make trouble for the government so that it can do its job transforming China into a great power.
(6) In the face of anti-Chinese provocations, wisdom and development are keys to counteracting such forces![43]

The results of a poll conducted in 2007 by Horizonkey suggest that Chinese Net users have become more inclined to espouse the resolution of international conflicts and issues through diplomatic cooperation. This contrasts with what they perceive as America's domestic democracy and international hegemony.

According to polls conducted by Horizonkey Group in 1999 and 2004, the Chinese people realize that the United States is supremely important for China's economy (figure 7.11).

However, a perception of American unilateralism sustains the Chinese perception that the United States is among the world's most unfriendly nations. In both 1999 and 2004, over 70 percent of respondents espoused the view that the United States was the world's most unfriendly country (figure 7.12).

That being said, another poll by the Horizonkey Group reveals that Chinese Net users nonetheless recognize the existence of many shared interests between the two countries, including those in the fields of health care (4.03 points), international crime and drug trafficking (4.01), the environment (3.77), terrorism (3.57), and East Asian economic growth (3.56). In fact, 67.4 percent of respondents believed that the United States and China both share common interests in promoting the economic growth of East Asia.

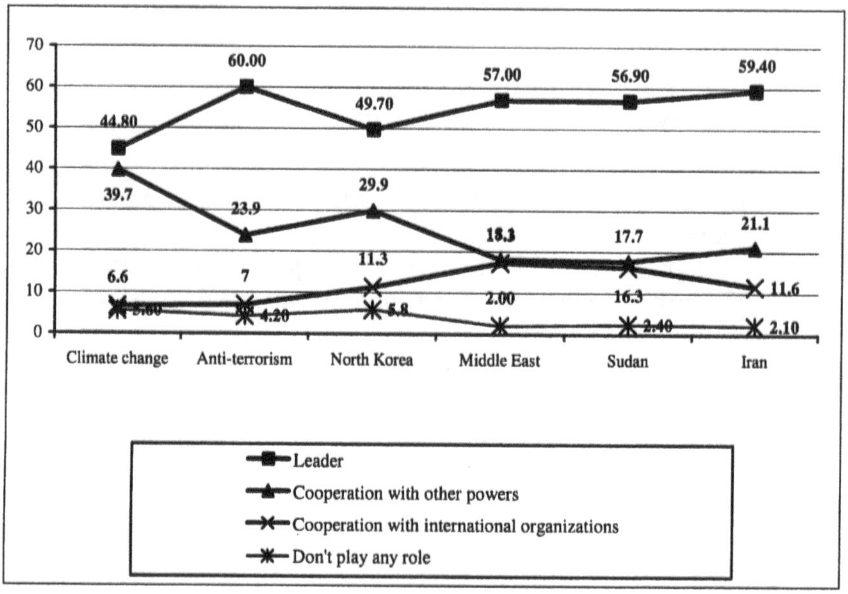

FIGURE 7.10
Chinese Views on China's Role in World Affairs

Source: "Diqu Daguo hai shi Chaoji Daguo? Zhongguo ren Yan Zhong de Weilai (Regional Power or Super Power? Chinese View Their Future)," Horizonkey Group, March 4, 2008, available at www.horizonkey.com/showart.asp?art_id=745&cat_id=6.

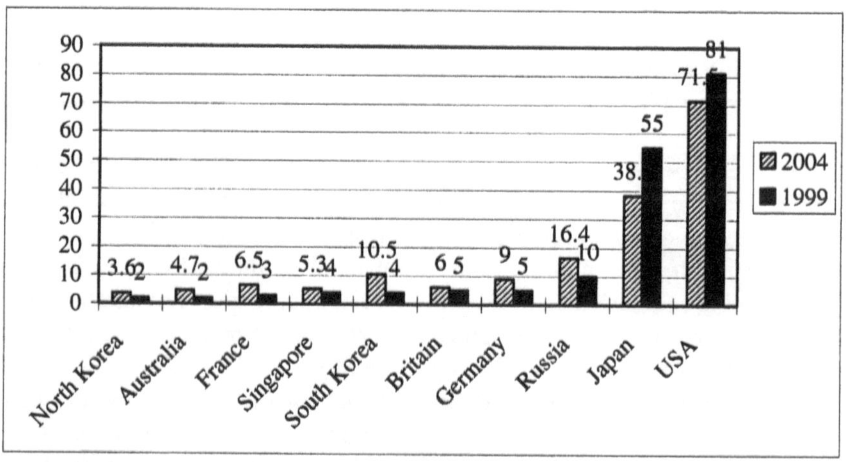

FIGURE 7.11
China's Most Important Economic Partner Nation (1999 and 2004)

Source: "Zhongguo ren kan Meiguo de Ganshou hen Fuzha (Chinese America Views are Complex)," Horizonkey Group, April 11, 2004, available at www.horizonkey.com/showart.asp?art_id=418&cat_id=6.

1999		2004	
Country	%	Country	%
USA	70.3	USA	74.4
Japan	52.6	Japan	66.3
Indonesia	16.3	India	9.2
India	15.4	Vietnam	6.0
Vietnam	6.1	Indonesia	4.9
Great Britain	5.9	Great Britain	4.9
Russia	2.0	North Korea	2.8
North Korea	2.0	Singapore	2.3
South Korea	1.4	Russia	2.3
Australia	1.4	South Korea	1.6

FIGURE 7.12
Chinese Perception of the World's Most Unfriendly Nation (1999, 2004)

Source: "Zhongguo ren kan Meiguo de Ganshou hen Fuzha (Chinese America Views are Complex)," Horizonkey Group, April 11, 2004, available at www.horizonkey.com/showart.asp?art_id=418&cat_id=6.

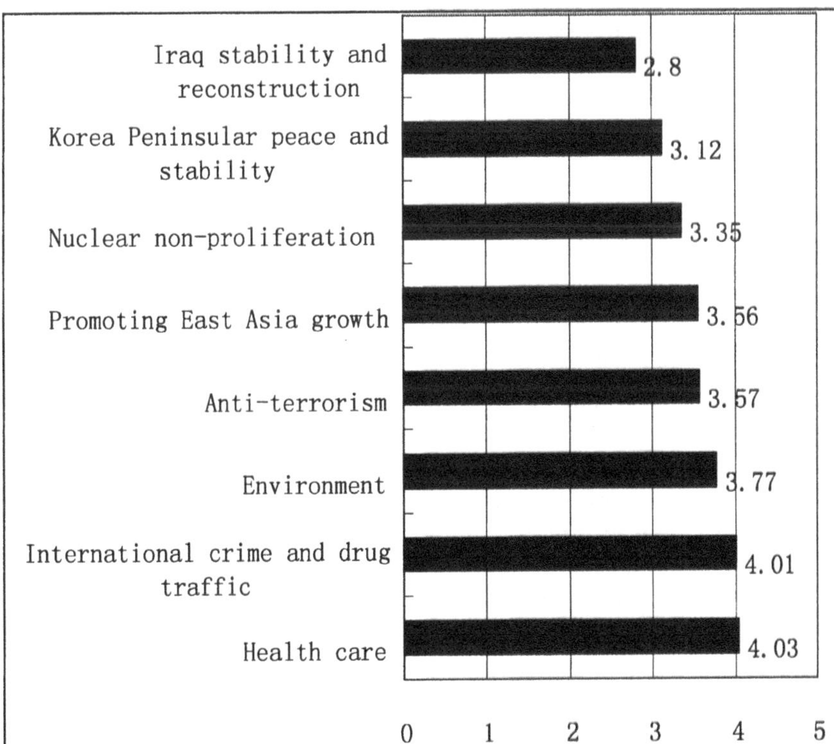

FIGURE 7.13
Perceptions of Sino-American Common Interests (2004)

Source: "Zhong Mei Guanxi jian ru Jiajin, dan Mingzhong Haogan du ren jiao di (Sino-America Relations Better Up, but Favorite Temperature Still Low)," Horizonkey Group, June 2005, available at www.horizonkey.com/showart.asp?art_id=480&cat_id=6.

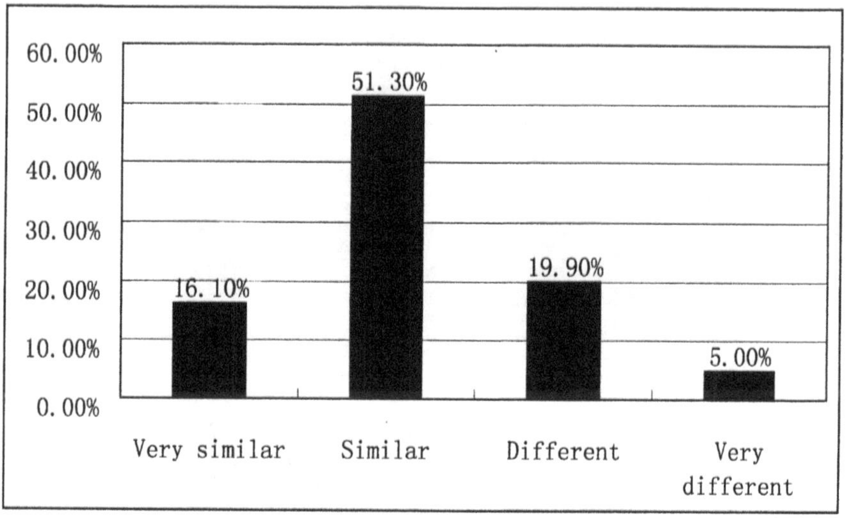

FIGURE 7.14
Sino-American Shared Interests in Promoting the East Asian Economic Growth (2004)
Source: "Zhong Mei Guanxi jian ru Jiajin, dan Mingzhong Haogan du ren jiao di (Sino-America Relations Better Up, but Favorite Temperature Still Low)," Horizonkey Group, June 2005, available at www.horizonkey.com/showart.asp?art_id=480&cat_id=6.

Conclusion

Ever since the American merchant ship *Empress of China* arrived in China in 1784, the United States has been regarded as an enigma from the Chinese perspective.[44] Almost all Chinese Net users actively embrace and utilize the dichotomy of two Americas: a "beautiful and advanced America," and a "hegemonic and imperialistic America." This dichotomy can be further broken down into three dominant views: a hatred toward America, a profound admiration of America, and an ambivalent stance toward America. That being said, online Chinese nationalism toward the United States is becoming increasingly rational, attenuating extreme views. The characterization of Chinese online nationalism is contingent upon the realm of the topic being discussed, and thus not as clear-cut as portrayed by the Western media.

The author acknowledges the limits of this study. The majority of Net users are well-educated and hail from a relatively privileged socioeconomic class. It is also important to note that many Net users refrain from presenting their views regarding the United States for various reasons, including a lack of will, interest, time, and/or energy. Therefore, this study only reflects one of many views concerning online Chinese nationalism vis à vis the United States. However, since the overwhelming majority of Net users are young, an analysis of their viewpoints toward the United States carries great import since they will ultimately shape China's future.

Notes

*The editing work of Christopher Lau, research assistant at International Relations Research Program at the Chinese University of Hong Kong in 2009, under the supervision of Simon Shen, is fully acknowledged.

1. Tiffany Danitz and Warren Strobel, "The Internet and Burma: A New Tool for Nonviolent Change," U.S. Institute for Peace Conference Paper.
2. Gary Wasserman, *Meiguo Zhengzhi Jichu* (Basics of American Politics), Chinese edition, trans. by Lu Zhen'guan (Beijing: Zhongguo Shehui Kexue Chubanshe, 1994), 131.
3. Wang Yiwei and Tang Xiaoshong, "Daguo Zhengzhi Beiju yu Guoji Guanxi Lilun de Beiju: Ping Mearsheimer de *Daguo Zhengzhi de Beiju* (Tragedy of Great Power Politics and Tragedy of International Relations Theory: On Mearsheimer's *Tragedy of Great Power Politics*)," John Mearsheimer, *Daguo Zhengzhi de Beiju* (Tragedy of Great Power Politics), Chinese edition, trans. by Wang Yiwei and Tang Xiaoshong (Shanghai: Shanghai People's Press, 2003), 28.
4. "Dui Meiguo gai Ai haishi heng? (Should We Love or Hate the United States?)," Sina Forum, Dec. 12, 2007, available at bbs.news.sina.com.cn/tableforum/App/view.php?bbsid=1&subid=0&fid=30885&tbid=4392.
5. Liu Yazhou, "Meiguo ZhenZheng de Kepa zhi chu zai Na'er (What is really so bad about the United States)," Qiangguo Luntan, Feb. 12, 2008, available at bbs.people.com.cn/postDetail.do?boardId=7&view=1&id=84729109.
6. Wang Xiaodong and Ma Mingjie, "Zhongguo Qingnian Kan Shijie (The World in the Eyes of Chinese Young People)," *China Youth Daily*, May 30, July 14, 1995
7. "Zhongguo Ren yan zhong de Shijie (The World in Chinese Eyes)," Horizonkey Poll, 1995, available at www. horizonkey.com.
8. Cao Xin, "Jin 10 Nian lai Zhong Mei Guanxi Diaocha zhong de Mingjian Renzhi (Non-Official Perspectives regarding Sino-America Relation Polls in the Past 10 Years)," *Newsweek*, May 2003.
9. Zhang Hui, "Zhongguo ren kan Meiguo: Shiyong Zhuyi + Bentuhua Qingjie (Chinese See the United States: Pragmatism plus Localization)," Horizonkey, Nov. 14, 2003, available at www.horizonkey.com/first/first_jb/594.htm.
10. Hoang Renwei, *Zhongguo Jueqi de Shijian yu Kongjian* (The Timing and Placing of China Rising) (Shanghai: Shanghai Social Science Academy Press, 2002).
11. See Maxine Thomas and Zhao Mei, eds., *Zhong Mei Changqi Duihua* (*Sino-America Long-term Dialogue*) (Beijing: China Social Science Press, 2001); Chen Shengluo, "Liang ge Meiguo: Zhongguo Daxuesheng de Meiguo guan (Two Americas: Chinese Undergraduates' America Views)," *China Youth Daily*, Sep. 5, 2006, available at www.youth.cn/qsnyj/200609/t20060905_377144.htm; and the polls of Horizonkey Group.
12. Chen Shengluo, "Liang ge Meiguo: Zhongguo Daxuesheng de Meiguo guan (Two Americas: Chinese Undergraduates' America Views)," *China Youth Daily*, Sep. 5, 2006, available at www.youth.cn/qsnyj/200609/t20060905_377144.htm.
13. "Dui Meiguo gai Ai haishi heng? (Should We Love or Hate the United States?)," Sina Forum, Dec. 12, 2007, available at bbs.news.sina.com.cn/tableforum/App/view.php?bbsid=1&subid=0&fid=30885&tbid=4392.

14. Zhao Mei, "Zhongguo kan Meiguo (Chinese View the United States)," in Maxine Thomas and Zhao Mei, eds., *Zhong Mei Changqi Duihua* (Sino-America Long-term Dialogue) (Beijing: China Social Science Press, 2001), 3–20.

15. Ling Chao, "Ba cheng Beijing ren Fandui Meiguo dui Yi Dongwu (80 percent Beijing Residents against United States Invading Iraq)," Horizonkey Group, Nov. 13, 2003, available at www.horizonkey.com/first/first_jb/555.htm.

16. "Fangan Zhanzheng: Zhongguo ren mei de shuo! (Anti-War: Chinese, absolutely!)," Horizonkey Group, Nov. 14, 2003, available at www.horizonkey.com/show art.asp?art_id=241&cat_id=6.

17. "Fangan Zhanzheng: Zhongguo ren mei de shuo! (Anti-War: Chinese, absolutely!)," Horizonkey Group, Nov. 14, 2003, available at www.horizonkey.com/show art.asp?art_id=241&cat_id=6.

18. "Zhongguo Gongzhong dui Mei Haogan da fu Tisheng (Chinese Public much more favorable towards the United States)," Horizonkey Group, Dec. 14, 2006, available at www.horizonkey.com/showart.asp?art_id=609&cat_id=6.

19. "Zhong Mei Guanxi jian ru Jiajin, dan Mingzhong Haogan du ren jiao di (Sino-America Relations Better Up, but Favorite Temperature still Low)," Horizonkey Group, June 2005, available at www.horizonkey.com/showart.asp?art_id=480&cat_id=6.

20. Horizonkey Group, "Meiguo ren ruhe kan Zhongguo (How American view China)," Feb. 14, 2001, available at www.horizonkey.com.

21. "Xiaoying Hao Shijian Jiekai Zhongguo Junshi Fangong de Xumu! (The *Kitty Hawk* Event Unveils the Prologue of Chinese Military Counterattack!)," Sina Forum, Dec. 11, 2007, available at bbs.news.sina.com.cn/tableforum/App/view.php?bbsid=1&subid=0&fid=30824&tbid=2016.

22. "China Demands Apology from CNN's Cafferty for 'Goons and Thugs' Comment," *Los Angeles Times*, April 16, 2008; "Thousands Protest CNN's Cafferty in Los Angeles," *Los Angeles Times*, April 20, 2008.

23. "China Demands Apology from CNN's Cafferty for 'Goons and Thugs' Comment," *Los Angeles Times*, April 16, 2008; "Thousands Protest CNN's Cafferty in Los Angeles," *Los Angeles Times*, April 20, 2008.

24. "Question the CNN's Qualification to Report the 2008 Beijing Olympic Games," Sina Forum, available at hi.news.sina.com.cn/news/anticnn/list.php?dpc=1, accessed on May 12, 2008.

25. "Chong CNN kan Meiguo dui Yazhou ge Guo Taidu (America's Attitudes toward Asian Countries from CNN's Perspective)," Tianya Community, March 22, 2008, available at cache.tianya.cn/publicforum/Content/worldlook/1/171692.shtml.

26. "Mei Xuanchuanbu Yanshu Chuli CNN Zhuchiren de Jueding (Decision to Punish Host of CNN Seriously by Department of Propaganda of the United States)," Tianya Community, April 21, 2008, available at cache.tianya.cn/publicforum/Content/free/1/1201469.shtml.

27. "Meiguo Juzhu Jinzhengli Qiangjing, Zhongguo ren ai hen liang nan quan (American Inhabitation Competitiveness Strong, Chinese Love and Hatred)," Horizonkey, Oct. 11, 2004, available at www.horizonkey.com/showart.asp?art_id=377&cat_id=6.

28. "Zhongguan Waiguo Jiaoyu Shitan Zhongguo Jiaoyu Gaige zhi lu (Observing Foreign Education System and Exploring Chinese Educational Reform)," Tianya Community, Nov. 10, 2007, available at cache.tianya.cn/publicforum/Content/free/1/1045155.shtml.

29. "Zhongguo si Jiaoyu de Kebo yu Meiguo si Jiaoyu de Kuanrong (The Acerbity of Chinese Style Education and Toleration of American Style Education)," Tianya Community, March 14, 2008, available at www.blog.qingyu.org/unkind-chinese-education.

30. "Zhaoyu Meiguo Jiaoyu (Encountering American Style of Education)," Tianya Community, Sep. 20, available at cache.tianya.cn/publicforum/Content/free/1/794673.shtml.

31. Chen Shengluo, "Liang ge Meiguo: Zhongguo Daxuesheng de Meiguo guan (Two Americas: Chinese Undergraduates' America Views)," *China Youth Daily*, Sep. 5, 2006, available at www.youth.cn/qsnyj/200609/t20060905_377144.htm.

32. "'Haigui' Guanghuan jian tui ('Haigui' Aura is Fading)," *China Youth Daily*, Oct. 11, 2003; "'Haigui' Chengwei Tuidong wo guo Jianshe Fazhan de Zhongyao Liliang ('Haigui' Become Important Force of Facilitating Our Construction and Development)," CCTV International News, 17:40, Nov. 17, 2003.

33. Zhang Hui, "Zhongguo ren kan Meiguo: Shiyong Zhuyi + Bentuhua Qingjie (Chinese See the United States: Pragmatism plus Localization)," Horizonkey, Nov. 14, 2003, available at www.horizonkey.com/first/first_jb/594.htm.

34. "Meiguo ren yong Yao Ming huan zhou le Shenme? (What are the American Gained from Accept Yao Ming?)," Bokee Net, March 21, 2008, available at hyh1973.bokee.com/viewdiary.25495989.html.

35. "Meiguo Xuesheng neng gan Shenme jiu gan Shenme ma? (Could American Students Do What They Want to Do?)," Tianya Community, Jan. 16, 2008, available at cache.tianya.cn/publicforum/Content/free/1/1099599.shtml.

36. Zhang Hui, "Zhongguo ren kan Meiguo: Shiyong Zhuyi + Bentuhua Qingjie (Chinese See the United States: Pragmatism plus Localization)," Horizonkey, Nov. 14, 2003, available at www.horizonkey.com/first/first_jb/594.htm.

37. Cao Xin, "Jin 10 Nian lai Zhong Mei Guanxi Diaocha zhong de Mingjian Renzhi (The Non-Official Perspectives in the Sino-America Relation Polls in the Past 10 Years)," *Newsweek*, May 2003.

38. Cao Xin, "Jin 10 Nian lai Zhong Mei Guanxi Diaocha zhong de Mingjian Renzhi (The Non-Official Perspectives in the Sino-America Relation Polls in the Past 10 Years)," *Newsweek*, May 2003.

39. Chen Shengluo, "Liang ge Meiguo: Zhongguo Daxuesheng de Meiguo guan (Two Americas: Chinese Undergraduates' America Views)," *China Youth Daily*, Sep. 5, 2006, available at www.youth.cn/qsnyj/200609/t20060905_377144.htm.

40. "Wangming Pupian Guanzhu Kongbu Huodong, Quanming Lianhe Fankong zhui shou Zhancheng (Net Users Pay Attention to Terrorist Activities, Favor Cooperative Anti-Terrorism)," Horizonkey Group, Nov. 30, 2002, available at www.horizonkey.com/showart.asp?art_id=216&cat_id=6.

41. "Wangming Pupian Guanzhu Kongbu Huodong, Quanming Lianhe Fankong zhui shou Zhancheng (Net Users Pay Attention to Terrorist Activities, Favor

Cooperative Anti-Terrorism)," Horizonkey Group, Nov. 30, 2002, available at www.horizonkey.com/showart.asp?art_id=216&cat_id=6.

42. Sheng Yu, "Guojia Yingxiang: Zhongguo ren zhui ren Zhongguo (National Image: Most Chinese Identify with China)," Horizonkey Group, Nov. 14, 2003, available at www.horizonkey.com/first/first_jb/593.htm.

43. "Yi Yu Jing Tan (One Wisdom Word)" (April 20), People Forum, April 20, 2008, available at bbs1.people.com.cn/postDetail.do?boardId=1&treeView=1&view=2&id=85466437.

44. Yang Yusheng, *Zhongguo ren de Meiguo Guan: Yi ge Lishi de Kaoca* (Chinese American View: A Historical Survey) (Shanghai: Fudan University Press, 1996).

III

THE SECOND LAYER: THE REST OF THE WORLD

8

Beyond Sino-ASEAN Relations

*Online Chinese Nationalism Toward Southeast Asia**

Chun Zhang

In general, China has long enjoyed positive relations with Southeast Asia. Through comments posted on Internet forums, many Chinese express their appreciation of the region's scenic natural landscape. Chinese Net users demonstrate a deeper knowledge of Southeast Asia compared to that of other regions. With this in mind, one posits that online Chinese nationalism toward Southeast Asia is largely characterized by rational and peaceful sentiments. Those who profess extreme views are most likely to be in the minority.

In order to adequately take into account the extensive array of Internet literature and the myriad of views expressed by Net users, the analysis has been divided into three sections. The first section presents a survey of Chinese Net users' general attitudes toward Southeast Asia, which are largely positive. Besides the significant number of ethnic Chinese living in the region, many Chinese are drawn to the tourist and business opportunities in the region. The following section discusses the political dimension of Chinese online nationalism toward Southeast Asia. Both cooperative and competitive dimensions of Chinese–Southeast Asian relations will be examined. Additional points of interest include the region's role in China's grand strategy (especially as it relates to the Sino-U.S. bilateral relationship) and whether or not China aims to transform the region into its "backyard." The chapter will also focus on several issues of contention that occasionally induces a departure from the status quo of favorable Chinese attitudes toward Southeast Asia. Topics that have been selected for analysis concern the South China Sea, Chinese overseas, and China-Vietnam relations.

The analysis is largely drawn from the views presented on several of the largest Internet communities, such as the Tianya Community, Qiangguo Forum, and Sina Forum. Such communities are characterized by their high registers and relatively comprehensive views. Despite their visibility, other sites such as the Tiexue (Iron and Blood Forum) and Military Forum of Sina.com are excluded either because of the extreme views presented or the forum's specificity with regard to its content.

A Favorable View of Southeast Asia

The rapid rise of the Chinese economy has contributed to the heightened popularity among Chinese of traveling abroad. Due to time, money, and cultural considerations, Southeast Asia is often considered the top destination of Chinese tourists. Not surprisingly, a significant share of Chinese online discussion vis-à-vis Southeast Asia is related to tourism. Successful online marketing campaigns have encouraged more and more Chinese to visit the region, further instilling a favorable image of the region.

It is estimated that more than 70 percent of all Internet comments concerning Southeast Asia are related to some aspect of tourism. Such comments range from those that only highlight the tourism of an individual state to those that portray the tourism of all ten ASEAN states. Illustrative pictures often accompany the comments. The following is a widely circulated text that presents a survey of what ASEAN states have to offer for tourists:

> Laos, a country sometimes overlooked by Chinese tourists***Brunei, a small state considered the "Venice of the East"***Indonesia, a country boasting the greatest tourist resources in the world***Malaysia, a country that is like a big leaf floating between the Pacific and Indian Oceans***Philippines, a country regarded as heavenly by Europeans and Americans***One can't possibly visit the more than 500 landscapes in Thailand***Cambodia's most famous tourist attraction is the Angor Wat, one of the most beautiful and mysterious historical sites in the world***Myanmar is the largest country within the Southeast Asian peninsula whose fastest rising industry is tourism***Singapore is a tropical island nation and is considered one of the world's garden cities***Halong Bay, located in northern Vietnam, is a World Heritage site and the perfect tourist destination***[1]

Chinese Net users' amity toward Southeast Asia not only stems from their admiration of Southeast Asia's offerings in the area of tourism, but also from traditional historical ties. Traditionally, Southeast Asia is known as "Nanyang" (Southern Ocean). From the Ming to Qing dynasty, a great number of

Chinese migrated to Nanyang, contributing greatly to the region's local economic development. Overseas Chinese not only enjoyed tremendous success in their new homes, but also made significant contributions to China (their original homeland). During World War II, Chinese overseas demonstrated their loyalty to China by donating resources and money toward the war effort. Scores of young overseas Chinese even returned to China and enlisted in the armed forces. Devotion to mainland China persisted even after the establishment of the People's Republic of China, as demonstrated by pro-Chinese unification sentiments of overseas Chinese.

Additionally, given the implementation of the "go global" strategy with regard to Chinese economic development, Southeast Asia has emerged as a preferred investment destination of Chinese companies. The following is the central argument in a paper entitled "Zuoba, Dao Dongmen Guojia Taojin Qu! (Go, Go to ASEAN States for Fossicking)":

> As the Chinese government encourages large nationalized companies to seize business opportunities abroad, Southeast Asia has been trying to recover from the shock of the 1997 Southeast Asian Financial crisis. The region is attempting to restore investor confidence, increase foreign direct investment, and enact policies conducive to business. Chinese companies should thus make good use of these reforms [in Southeast Asia] and invest in local companies there. Companies from Taiwan, Hong Kong, Singapore, Europe, and the United States have already engaged the region. Chinese companies should emulate the structural adjustment framework of Japanese companies in order to explore Southeast Asian markets.

The author of the paper also suggests a three-pronged strategy to implement the "go-global" strategy: (1) highlight one's advantages while attacking others' disadvantages, (2) meet an opponent head-on in home court, (3) challenge an opponent in an away court, and (4) diversify business strategies.[2]

On the topic of business, many inquiries are put forth online regarding information concerning the business environment, investment laws, and investment feasibilities of Southeast Asia. Official, civil, and nonofficial websites all provide useful information.[3]

Given the beauty of its landscapes and the existence of loyal Chinese overseas and business opportunities in the region, Chinese Net users profess a rational and optimistic attitude toward Southeast Asia. A widely circulated paper written by a Japanese scholar echoes this sentiment:

> China-ASEAN relations are thought to be "the best in the past 50 years." Due to a positive attitude on the part of the Chinese, this relationship is among the friendliest of bilateral relations between international entities. China was the first ASEAN dialogue partner to join the Treaty of Amity and Cooperation in

> Southeast Asia and to establish the strategic partnership for peace and prosperity with ASEAN. China was the first to initiate the process of establishing a free trade regime with ASEAN. Both parties have already signed the Framework Agreement on Comprehensive Economic Cooperation. China also exercises soft power in the region. It is reported that in Southeast Asia, 800 high schools teach Chinese while only 250 schools teach Japanese. Under the Declaration on the Conduct of Parties in the South China Sea, China and ASEAN have agreed to academic and cultural cooperation through initiatives such as teacher and student exchanges and joint research projects. China has enrolled more ASEAN students in its schools than Japan has.[4]

There are those who are simply unaware of China's close ties with Southeast Asia, as demonstrated by online users who criticize the Chinese government's disproportionate attention toward industrialized countries compared with that directed toward geographical neighbors. However, fellow users are often quick to expose the ignorance of such a presumption.[5]

Despite the overall friendly characterization of the China-ASEAN relationship, negative sentiments are still expressed. For example, when natural calamities abounded in spring 2008 in mainland China, some Chinese Net users questioned why Southeast Asian countries failed to offer and provide assistance for China (since China had extended a helping hand in the aftermath of the 2005 tsunami).[6] Although China and ASEAN have made considerable progress in forging economic cooperation, many are concerned about the deleterious impact of economic agreements on agricultural production in China's Hainan Province. Such concerns are expressed in a text entitled "Dongmeng Nongcanpin Jingru Zhongguo ling Guanshui, Daju Jiya Hainan Nongcanpin (ASEAN Agricultural Products Entering China with Free Tariffs Greatly Suppress Production of Agricultural Products)." According to the author, special fruits and vegetables flourish in Hainan Province due to its unique geography. However, the climate of Hainan is similar to that of Thailand and Vietnam, where such fruits and vegetables also flourish. The author thus warns that the importation of tariff-free agricultural products from Thailand and Vietnam would have devastating ramifications on the agricultural industry of Hainan. To compound Hainan's predicament, agricultural industries in Thailand and Vietnam are more cost-effective due to relatively cheap production, packaging, preservation, and shipping costs. On a more radical note, some believe that the province of Guangxi should eschew business relations with countries with the same level of development as Guangxi (i.e., Vietnam, Myanmar, Laos) and only trade with wealthier Chinese provinces such as Guangdong. According to these protectionists, Guangxi's economic growth can only be ensured by the implementation of this policy.[7]

ASEAN's Strategic Position

Chinese Net users tend to speculate on how Southeast Asia factors into China's grand long-term strategy. Given China's astronomical rise and ambitious goals, recurring questions that come to mind include: Will Southeast Asia become China's backyard? Will China engage in competition with other countries for influence in this region? If so, what kinds of competition will occur and what are the prospects for mediation and resolution? The majority of Net users exercise rational restraint in grappling with such questions. Indeed, many Chinese Net users worry about the possible competition posed by the United States, Japan, and India in the region. That being said, most prefer to abide by the five principles of peaceful coexistence. Supporting the ASEAN+3 framework, Chinese Net users hope China will pursue peaceful development objectives and abstain from establishing an exclusive sphere of influence in the region.

Nonetheless, a vocal online minority insists that Southeast Asia must become China's geopolitical backyard. They reason that if China aspires to be a veritable world power, it needs to have its own sphere of influence. The following is a presentation of their key arguments:

> Southeast Asia is in China's backyard since there are no other great powers in that region. China should make good use of its vast resources and development potential. Otherwise, China will not be able to become world power. The presence of Overseas Chinese in the region bolsters China's claim to the region. . . . With regards to the struggle for becoming the number one country in the world, China has not acquired enough power to compete with the United States. However, the United States does not exercise supreme influence in every corner of the globe. China can thus concentrate on consolidating its power in Southeast Asia and use this as leverage vis-à-vis the United States. Once Chinese influence attains a certain level, the United States may withdraw altogether from Southeast Asia. The establishment of a sphere of influence thus compels great powers to recognize our status as a regional power.[8]

The following proclamation is even more radical. The author outlines steps China must enact to ensure its long-term interests and existence:

> First, transform Taiwan into a base for reoccupying the South China Sea. Southeast Asian states will no longer have access to capital and therefore be unable to counteract China's efforts. Oil storage capabilities would exist for at least 20–50 years. Second, establish Southeast Asia as China's backyard region so that it can provide colored metal storage for 50 years. Third, reoccupy South Tibet and the Indian Ocean. Then take over Africa, a region that can provide resources for 300–500 years.[9]

It is important to reiterate that such aggressive views are in the minority. Many Chinese Net users are aware of the "pay tribute system" that existed historically between China and the states of Southeast Asia but realize that such a system will most likely never be revived. From a realist perspective, China has never been the "center" of Asia, let alone the "hegemon" of Asia. China may overtake Japan economically in the future, but will only attain power parity with Japan at best. The country that exercises hegemonic influence in Asia is neither China nor Japan, but the United States. Both the United States and Europe have historically exerted tremendous influences that are deeply embedded in the fabric of Southeast Asian states. In light of this, China must consider realistic and innovative possibilities for cultivating positive relations with its Asian neighbors.[10]

Chinese Net users acknowledge the concerns raised by ASEAN of being marginalized given China's rising influence in the region. They therefore pay careful attention to ASEAN's initiatives to balance China by welcoming other powers into the region.

In ASEAN's annual regional forum held in June 2004, ASEAN proposed a new geopolitical strategic initiative dubbed by Japanese media "concentric diplomacy." Similar to Great Britain's "three circles diplomacy" following the Cold War, the "concentric diplomacy" model categorizes foreign powers based on their importance to Southeast Asia. In this model, ASEAN is at the "center," Japan, China, and South Korea comprise the "inner circle," while India, Russia, and Australia exist in the "outer circle." Chinese net users view ASEAN's adoption of this model as a manifestation of ASEAN's growing desire to firmly establish its geopolitical position within the Asian regional context.[11]

The rivalry between Japan and China for influence in the region will first be examined. China was the first country that proposed a free trade area with ASEAN. In 2001, China also proposed the strengthening of its bilateral relations with ASEAN in both the political and economic realms. The day after China's announcement, Japan proposed a free trade agreement with ASEAN. In attempting to understand Japan's actions, Chinese Net users provide the following analysis:

> Southeast Asia is a key component of Japan's international business and desire to balance China in Asia. Japan ultimately wants to integrate Southeast Asia into its business and commercial circles. Politically, Japan needs ASEAN's support in order to realize its great power status. However, Japan's historical conduct as well as its continued friendship with the United States has hindered the progression of its relations with Southeast Asian countries.

Although the aggregate volume of trade between China and ASEAN is lower than that of Japan and ASEAN, the current rate of growth in the former partnership is much higher than that of the latter. Moreover, China has signed the Treaty of Amity and Cooperation in Southeast Asia, a positive step toward the building of the China-ASEAN Free Trade Zone.[12]

China also has to contend with another power in the struggle for influence in the region: the United States. Chinese Net users have closely followed American assessments of China's role in Southeast Asia. For example, Net users often use the following quote by American China specialist Joshua Kurlantzick to justify China's supposed claim to Southeast Asia:

> Perhaps, as a young United States once did in the Western Hemisphere, China could make the region its own—a Chinese Monroe Doctrine for Southeast Asia would make Beijing the major influence over regional affairs and reduce US presence and alliances in the region.[13]

Chinese Net users have also noted a report published by the U.S. Congressional Research Service, which concluded that China has achieved significant success in Southeast Asia. The report also suggests that Beijing has effectively assuaged concerns of Chinese hegemony.[14]

Not surprisingly, American military exercises conducted in Southeast Asia often capture the interest of Net users. Such users speculate on the reasons behind the continued U.S. military presence in the region, and have concluded that such military exercises have three aims. The first is that the United States is perpetually on the lookout for new places to establish military bases. After all, in the 1960s and 1970s, the United States stationed armed forces in Vietnam and the Philippines (such forces have since withdrawn). The second aim is to strengthen joint military exercises with regional states. The third objective is to station military servicemen in the region under the pretext of providing anti-terrorism support. There are many Chinese who believe that the U.S. interest in Southeast Asia lies with the region's strategic position in enabling the United States to simultaneously intervene in Middle Eastern affairs while keeping an eye on China.[15]

In order to balance the rise of China, many ASEAN states have also begun to build up their relations with India, mainly through the strengthening of economic ties. Singapore was the first country to initiate such ties with India, assisting the island nation's "co-development with a rising India." Interestingly, ASEAN's "look west" policy overlaps with India's "look east" policy, a dynamic which has great implications for ASEAN-Indian relations.[16]

Lastly, ASEAN has vigorously promoted community building among its members in order to prevent its marginalization by foreign powers. At the end of 2007, ASEAN states ratified the ASEAN Charter, which established ASEAN as a legal entity. To ensure durable peace, stability, and shared prosperity in the region, members agreed to three fundamental pillars: political and security cooperation, economic cooperation, and sociocultural cooperation. Member states also ascribed to ASEAN's Vision 2020, which called for the establishment of an ASEAN Economic Community. Such a community would guarantee the free flow of goods, services, and investment in the region. Equitable economic development, the reduction of poverty, and the elimination of socioeconomic disparities were also goals to be attained by the year 2020.[17] Recently, ASEAN economic ministers have agreed to establish the ASEAN Economic Community by 2015.[18]

Given the competing interests of China, Japan, the United States, India, and other states in Southeast Asia, the question inevitably arises: How can peace be preserved? The majority of Chinese Net users both dismiss the transformation of Southeast Asia as China's backyard and war as means of settling competing interests. Instead, online communities often propose the heightened leadership and strengthening of ASEAN to preserve regional stability. Many are optimistic of the constructive influence ASEAN+3 could play in terms of its aim of building a large common market (like that of the European Union) through peaceful negotiations among ASEAN member states and key foreign players. Despite Russian and American opposition and EU and Japanese suspicion of ASEAN's ambitious plans, Chinese Net users have also suggested that ASEAN+3 become ASEAN+4 by enlarging the bloc to include India.[19]

Issues of Particular Contention

Up to this point, Chinese Net users' online nationalism as manifest toward Southeast Asia appears to be rather moderate and rational. That being said, the next section of this chapter will present issues that often incite more aggressive Chinese attitudes toward the region. The issues that will be discussed include those concerning the South China Sea, overseas Chinese, and China-Vietnam relations.

The South China Sea

Historically, the South China Sea has fallen under Chinese jurisdiction. That being said, from the 1960s onward, several islands of the Nansha archipelago have been occupied by neighboring countries. Until 1991, China controlled only Taiping Island (Itu Aba Island) and six reefs while forty-four other islands were occupied by Vietnam, the Philippines, and Malaysia. Those three countries, along with Brunei and Indonesia, granted sovereignty to several of the forty-four islands. Many Chinese Net users regard Vietnam as

China's most aggressive and primary adversary due to Vietnam's occupation of twenty-nine islands and exploitation of already more than thirty million tons of oil in the region.[20] In attempting to understand the historical and contemporary geopolitical realities of the South China Sea, Chinese Net users attribute the influence of four factors.[21]

First, Chinese Net users are aware of the Chinese navy's limited operation capabilities in the deep sea due to Chinese warships' very limited tonnage. The unsustainable ratio of warships–logistics ships also undermines the power projection capacity of the Chinese navy to effectively manage distant islands scattered around the South China Sea.[22]

Partly due to the Chinese navy's inability to effectively monitor the entire region, the geopolitical significance of the sea has elicited continued American and Japanese interest. To maintain its favorable strategic positions in the Asia-Pacific region and counteract the rise of China, both the United States and Japan actively seek the support of ASEAN states. Struggling to become Asia's *de facto* leader, Japan has attached great importance to its relationship with ASEAN member states. The diplomatic influence of Japan has consequently shifted to Southeast Asia. Confronted with both the United States and Japan's maneuverings, ASEAN states pursue a balancing strategy, employing both the "U.S. card" and "Japanese card" to advance their own interests in the South China Sea.

The actions of the Republic of China also contribute to the complex dynamics at play in the region. In order to maintain and expand its "international space," Taiwan has strengthened its economic ties with ASEAN, indirectly exercising leverage vis-à-vis China through ASEAN. Likewise, the Taiwan issue has provided ASEAN with leverage to effectively counteract China's claims of sovereignty over islands in the South China Sea.

As a collective entity, ASEAN has called for negotiations with China regarding disputes in the South China Sea. However, the declaration of ASEAN in 1994 that ASEAN members will only negotiate with external countries on a consensus basis (repudiation of bilateral negotiations) has and continues to exacerbate difficulties in resolving disputes.[23]

With the above factors in mind, Chinese Net users have proposed various suggestions. The following summarizes a widely held approach:

> For a rapidly developing China, the South China Sea is a bastion of hope, a region of tremendous economic and military importance. However, from a realist perspective, the resolution of the complex Nansha Islands dispute will be gradual. It is crucial to recognize the following five relationships: (1) the relationship between the Ocean and China's power projection capabilities, (2) the relationship between the reunification of Taiwan and the resolution of the Nansha Islands dispute, (3) the relationship between sovereignty and

national interests, (4) the relationship between China-ASEAN relations and China-Southeast state bilateral relations, and (5) the relationship between Great Powers and China-ASEAN relations.[24]

Another widely circulated proposal suggests that China "establish itself in Myanmar, support Thailand, outperform Indonesia, reel in Malaysia, isolate Singapore, punish Vietnam and the Philippines, while entertaining cordial relations with Laos, Cambodia, and Brunei." The author also proposes six guiding points toward the implementation of this rather radical strategy:

(1) Make use of Confucianism;
(2) Make use of Buddhism;
(3) Explore the role of overseas Chinese;
(4) Seek international cooperation in the development of the Greater Mekong Delta;
(5) Oppose Japan;
(6) Weaken dependency on the U.S. dollar and expand the market of Chinese *renminbi*.

The author does not advocate that Southeast China become China's backyard, and instead wishes that the relationship between China and Southeast Asia be distinguished by principles of goodwill, mutual cooperation, and dialogue.[25]

Overseas Chinese

The largest, arguably most successful, subcategory of overseas Chinese is the Southeast Asian overseas Chinese. Ten million Chinese overseas reside in Indonesia, a country whose total population is two hundred million, six million in Malaysia with twenty-five million total, and nine million in Thailand with sixty-five million total. In Singapore, a city-state with a population of three million, more than 90 percent of residents are of Chinese descent. There are one million overseas Chinese in the Philippines, a country with a total population of more than ninety million. However, more than 20 percent of Filipinos are of Chinese descent. Overseas Chinese in Southeast Asia are often highly successful in business and have dominated the economy. In Malaysia, 70 percent of small and mid-sized manufacturing, infrastructure-building, and service industries are controlled by overseas Chinese. When China liberalized and reformed its economic policies from 1979 to 1988, overseas Chinese investors flocked to Southeast Asia.

Malaysian-Chinese and Singaporean-Chinese are held in high social regard. However, a differentiation between "traditional Chinese" and "Straits

Chinese" exists in both countries. The "traditional Chinese" are often the owners of businesses. They remember their motherland (China) and adhere to the Chinese way of life. On the other hand, "Straits Chinese" often work for Western capitalists. They have abandoned their Chinese identity, harboring no affinity whatsoever toward China. In general, the majority of Malaysian-Chinese are "traditional Chinese," while the majority of Singaporean-Chinese are "Straits Chinese."[26] Malayan youths tend to identify more with China than their Singaporean counterparts. Many Malaysian-Chinese are reluctant to marry indigenous Malayans because the majority of Malayans are Muslim, and thus they would have to convert to Islam and abandon traditional Chinese culture.

On the other hand, overseas Chinese have been subject to rampant discrimination and marginalization in countries such as Indonesia. After the 1965 coup, the Indonesian government (with the support of the U.S. government) precipitated an anti-Indonesian-Chinese agitation that resulted in the deaths of nearly one million Indonesian-Chinese. In 1998, civil unrest in Indonesia was again directed against Indonesian-Chinese. After the eruption of the two agitations, Indonesian-Chinese hired armed patrols to protect their property and began to divest investments from Indonesia. For decades, discrimination against those of Chinese descent was sanctioned by the Indonesian government, which passed laws making it illegal to speak Chinese and prevented individuals with Chinese names from attending school. However, many Indonesian-Chinese secretly preserved their Chinese heritage in the confines of their own homes.[27] In 1958, in accordance with a treaty between the Chinese and Indonesian governments, Indonesian-Chinese lost their dual citizenship and hence a significant dimension of their identity. It was not until July 2006 that Indonesia enacted new nationality laws that abolished discrimination against Indonesian-Chinese.[28] The low social status accorded to some Southeast Asian overseas Chinese has contributed to such individuals' growing willingness to embrace their Chinese identity. However, their conceptualization of this identity is rather complex. In a conversation with a Chinese mainland individual, an Indonesian-Chinese stated, "You, Chinese are the grandsons of Dragon, the sons of Dragon's son; we, Southeast Asian Chinese, are the grandsons of Dragon, but the sons of Dragon's daughter."[29]

Not surprisingly, many Chinese are greatly offended by the discriminatory practices against overseas Chinese in countries such as Indonesia. In response to anti-Malaysian-Chinese agitations, a Net user posted the following remark:

> I call on you not to visit the country of Malaysia and refrain from contributing to their economy since Malaysia is discriminating against people of Chinese

> heritage. The Malaysian economy is dependent to a great extent on local Chinese; the indigenous competition is insignificant. If Malaysia did not have any Chinese, then it would be like any African country. The local government limits the number of Chinese officials, and the wealth generated by local Chinese is used to feed indigenous Malayans. Malayan-Chinese will soon be subject to the same ill-treatment endured by Indonesian-Chinese.[30]

Other Net users have noted that in Malaysia and Indonesia, Chinese inhabitants' aversion to politics is inimical to their well-being. Overseas Chinese thus become the defenseless targets of virulent, populist political sentiments. Many Chinese Net users continue to worry about the plight of Chinese in Southeast Asia, and express their wish that those in charge of Chinese foreign policy begin to address the issue.[31] Most believe that as China rises and its influence expands, the lives of overseas Chinese in Southeast Asia will greatly improve.[32] Rather than resorting to aggressive and violent tactics, the online Chinese community believes that China's development will ultimately lead to the amelioration of the status of Southeast Asians.

China-Vietnam Relations

Whenever the topic of Sino-Vietnamese relations is raised, most Chinese bitterly recall how Vietnam "bit the hand that fed them" at the conclusion of the Vietnam War, even though China had magnanimously aided the country during the war. Many Net users profess the utmost hatred toward Vietnam for the country's perceived betrayal, yet there are those who calmly (and rationally) analyze the reasons behind Vietnam's unpredictable behavior. A widely circulated analysis posits:

> The fundamental cause of the ebb and flow of Sino-Vietnamese relations is the restrictive ramifications resulting from the China-US-Soviet Union triangular relationship that existed during the Cold War. Sino-Vietnamese relations were contingent upon developments in Sino-US and Sino-USSR relations. When Sino-USSR relations dipped, China provided military and political assistance for Vietnam and hoped that Vietnam would not support the USSR in return. When Sino-US relationship improved, China asked Vietnam to come aboard. Vietnam could not tolerate this kind of erratic behavior on the part of great powers. Given the fluctuating dynamics of China-US-USSR relations, China's attitude towards Indochina changed from that of armed liberation to peaceful co-existence, a policy that ultimately alienated Vietnam to the point that it finally resorted to warfare.[33]

The South China Sea dispute between China and Vietnam has engrossed the interest of many Chinese Net users recently. Some Net users argue that given Vietnam's occupation of vast area and resource interests in the sea,

Vietnam has acquired the "sugarplum of aggression" and is harvesting its "fame and gain." The following summarizes their views on Vietnam:

> Vietnam's oceanic claim in the South China Sea is 3-4X the size of its own land acreage. This fact allows Vietnam to pursue its dream of becoming a "sea power." Second, since Vietnam also aspires to become an "oil power," it is greatly interested in the oil potential of the South China Sea. To realize these two dreams, Vietnam employs the military strategy of "ensuring the fruit of its aggression," the political strategy of containing China by using "China's strategy limitations and anti-Chinese international sentiments," and the ideological strategy of encouraging the Vietnamese people to accept the "concepts of sea and resource power."[34]

A few Net users, infuriated by Vietnam's uncompromising stances regarding South China Sea disputes, advocate the following:

> The Chinese army should mobilize for war. High level decision-making officials should prepare the right policy tools for dealing with Vietnam (especially in war time). Officials should draft war plans that will back islands and realize other strategic goals. High level officials should establish the goal of destroying 20–80% of the Vietnamese Navy. These steps will help guarantee the continued security of China's sea lanes.[35]

The irrationality of this view is dismissed by the majority of Chinese Net users. Instead, most urge China to continue its peaceful rise and development and make use of all opportunities to facilitate open dialogue with Vietnam. The prevailing sentiment is as summarized:

> There are four reasons why Vietnam chooses to provoke China. First, China's current priority is to develop its economy, and Vietnam knows that China will not declare war on Vietnam. Second, China continues to face the extremely difficult problem of how to unify Taiwan with the Mainland, which detracts attention away from Vietnam's expansion in the South China Sea. Third, China's most important bilateral relationship is that with the United States, and China does not have the energy or will to wage war with Vietnam. Lastly, since China is not yet a developed power, Vietnam is using this window of opportunity to build up its influence in the region. Once China attains world power status, Vietnam's prospects will be dashed.[36]

Under such circumstances, Net users have articulated what they believe are China's options:[37]

(1) *Toleration.* Since China is undergoing its renaissance, a war with Vietnam can be ruled out. However, preparations can still be made in the event of a war.

(2) *Deterrence.* The Chinese government will never choose to wage war with Vietnam. However, the government may deter and warn Vietnam of actions that might lead it to reconsider.
(3) *Coercion.* China has a wide array of coercive tools in its arsenal. For example, China could take advantage of Vietnam-U.S. differences and furnish arms for Laos and Cambodia.

Conclusion

Chinese online nationalism as directed toward Southeast Asia can thus be examined on three different levels. At the most general level, Chinese Net users harbor highly positive sentiments toward the region due to the abundance of tourism and business-related opportunities. Furthermore, many feel an affinity toward overseas Chinese living in the region. The middle level concerns Southeast Asia's role within China's grand strategy. The majority of Net users do not support the suggestion that Southeast Asia become China's backyard. Instead, they advocate for the continued progression of relations with states in the region based on the five principles of peaceful coexistence and development. The final level of analysis takes into consideration issues of special contention between China and Southeast Asian states. Certainly, the existence of such disputes complicates bilateral and multilateral relations. However, given the repudiation of extreme views by mainstream online users, Chinese online nationalism toward Southeast Asia is rather peaceful and rational in nature.

It is important to note that this chapter only reflects the attitudes of some Chinese Net users and not those of the Chinese populace at large. This chapter is thus representative of a rather privileged demographic of Chinese society. Net users who promulgate their views often possess some level of education, are financially stable, and are genuinely interested in Southeast Asia. The analysis found in this chapter therefore should be regarded with Net users' background in mind.

Notes

*The editing work of Christopher Lau, research assistant at International Relations Research Program at the Chinese University of Hong Kong in 2009, under the supervision of Simon Shen, is fully acknowledged.

1. "Xin xin zou zhou you bian Dongmeng shi guo (Tour of Ten ASEAN States)," Tianya Community, Oct. 26, 2007, cache.tianya.cn/publicforum/Content/travel/1/137053.shtml.

2. "Zuoba, Dao Dongmeng Guojia Taojin Qu! (Go, Go to ASEAN States for Fossicking)," Tianya Community, Jan. 27, 2003, cache.tianya.cn/publicforum/Content/develop/1/8435.shtml.

3. The most prominent investment website is the China-ASEAN Free Trade Area website, see www.cafta.org.cn.

4. "Ri Kan Jinji xue ren: Zhongguo yu Dongmeng guanxi '50 nian lai zhui hao' (Japanese Journal: China-ASEAN Relationship 'The Best Time in Recent 50 Years')," The People Website, Dec. 28, 2007, bbs.people.com.cn/postDetail.do?view=1&id=84375283&bid=6.

5. "Zhengfu gai Tiaozhen Waijiao de Zhongdian Fangxiang le (It's Time for Government to Re-Orient the Direction of Foreign Policy)," Tianya Community, Apr. 26, 2008, cache.tianya.cn/publicforum/Content/worldlook/1/176660.shtml.

6. The People Forum, Feb. 1, 2008, bbs.people.com.cn/postDetail.do?view=1&id=84656865&bid=1.

7. "Dongmeng Nongcanpin Jingru Zhongguo ling Guanshui, Daju Jiya Hainan Nongcanpin (ASEAN Agricultural Products Entering China subject to Free Tariffs greatly suppress Hainan Province's Agricultural Products)," Tianya Community, Sep. 28, 2007, cache.tianya.cn/publicforum/Content/hn/1/33126.shtml.

8. "Zhongguo zhen ba Tianxia zi Jingdian Chelue (The Classical Tactics of China Struggling for World Power Position)," Tom Military Forum, Sep. 27, 2004, army.news.tom.com/1019/1351/2004/9/27-27622.html.

9. "Ganjue Zhongguo de Gaige Kaifang Keneng Qiangongjingqi (Afraid that all of China's Reforms and Opening Efforts Have Been Wasted)," Linux Eden Forum, Feb. 29, 2008, www.linuxeden.com/forum/thread-162206-1-10.html.

10. "Tansuo Zhongguo yu Yazhou Guanxi de xin Moshi (Exploring New Models for China-Southeast Asia Relationship)," Jinhua Forum, Jan. 2, 2004, bbs.qianlong.com/redirect.php?tid=495456&goto=lastpost.

11. "Dongmeng Haipa Zhongguo Yingxiangli (ASEAN Worried about China's Influence)," Tianya Community, June 16, 2004, cache.tianya.cn/publicforum/Content/worldlook/1/32219.shtml.

12. "Dongmeng Tuidong Yazhou Jinji Gongtongti, Zhongguo Shenji Chenwei lao Dage (ASEAN Pushing for Asian Economic Community, China Become the Big Brother)," People Net, Jan. 16, 2007, bbs.people.com.cn/postDetail.do?view=1&id=1339023&bid=1; "Dongmeng Ziyou Maoyiqu—Zhong Ri xin Taipinyang Zhanzheng de jue Zhangchang)," Tianya Community, Feb. 28, 2004, cache.tianya.cn/publicforum/Content/worldlook/1/13704.shtml.

13. Joshua Kurlantzick, "A Newly Diplomatic China Courts the World," *Asia Sentinel*, May 25, 2007, www.asiasentinel.com/index.php?option=com_content&task=view&id=507&Itemid=34.

14. "Meiguo Wangzhan pin Zhongguo zai Dongnanya Yingxiangli, Yunan ren zui Kongbu Zhongguo (American Website Assesses Chinese Influence in Southeast Asia, Vietnam worried most about China)," Bokee Forum, Jan. 15, 2008, y326.bokee.com/viewdiary.23399035.html.

15. "Chong 'Kelake' Yanxi kan Mei zai Dongnanya de Junshi Cunzai (Observing American Military Existence in Southeast Asia through Military Exercises),"

Chinanet, July 31, 2003, www.china.com.cn/international/txt/2003-07/31/content_5376379.htm.

16. "Dongmeng Kaipi Yindu Shichang Zhiheng Zhongguo (ASEAN Explore Indian Market in order to Balance China)," Tianya Community, Dec. 2, 2004, cache.tianya.cn/publicforum/Content/worldlook/1/46389.shtml.

17. ASEAN, *Declaration of ASEAN Concord II* (*Bali Concord II*), Oct. 8, 2003, www.aseansec.org/15159.htm.

18. "The Politics of Speed: An ASEAN Community by 2015?," *The Nation*, Dec. 12, 2006, www.nationmultimedia.com/2006/11/27/opinion/opinion_30020037.php.

19. "Dongmeng 10+3 Fangshi Huajie Dongya Maodun? Keneng ma? (Resolve East Asian Differences by establishing ASEAN 10+3? Is it Possible?)," Tianya Community, Feb. 22, 2005, cache.tianya.cn/publicforum/Content/worldlook/1/54545.shtml.

20. "Shui shi Tiaozhan Zhongguo Nansha Zhuquan zui Weixian de Diren (Who is the Most Dangerous Adversary of Challenging China's Nansha Islands Sovereignty)?" China Zangnan Net, Jan. 2, 2008, www.chinazangnan.com//show.aspx?id=1851&cid=19; "Woguo Weilai Nansha Zhengduzhan de si ge zhu Zhanchang (Four Main Battlefields of Future Struggles for Nansha Islands)!" Blogs of 163 Net, March 16, 2008, helaibiao.blog.163.com/blog/static/20774732008216125321l.

21. "Dongmeng Guojia Lianshou shi Zhongguo Nanhai Xinshi Yanjun (ASEAN Members Joint Hand Complexed the South China Sea Issue)," Mil.Eastday, Feb. 21, 2008, mil.eastday.com/m/20080221/u1a3417581.html.

22. "Zhngguo Nanhai Wenti nan yi Jiaijue shou yin: Haijun Yuanyang zhan li Youxian Nansha Wenti de Chengying (The Main Reasons Why the South China Sea Issue Can't be Resolved: Navy Limited Operation Capability Is the Most Important One)," Sina Military, Jan. 21, 2008, mil.news.sina.com.cn/p/2008-01-21/0840482062.html.

23. "Dongmeng Guojia Lianshou shi Zhongguo Nanhai Xinshi Yanjun."

24. "Zhngguo Nanhai Wenti nan yi Jiaijue shou yin."

25. "Zhongguo Zhanlue."

26. "Huaren zai Nanyang (Chinese Who Live in Southeast Asian States)," Tianya Community, Feb. 16, 2008, cache.tianya.cn/publicforum/Content/worldlook/1/169017.shtml.

27. "Malaixiya Huaren zhuanti (The Malaysian-Chinese Special Issue)," Tianya Community, Oct. 3, 2003, cache.tianya.cn/publicforum/Content/worldlook/1/3571.shtml.

28. "Malaixiya Huaren zhuanti (The Malaysian-Chinese Special Issue)."

29. "Wo zai Ouzhou shuo Jiechu dao de Taiwan ren, Xianggang ren, Dongnanya Huaren (The Taiwanese, Hongkongese, and the Southeast Asian Chinese I Met in Europe)," Tianya Community, Sep. 11, 2003, cache.tianya.cn/publicforum/Content/worldlook/1/2578.shtml.

30. "Guanfu dao Huaren Zhunyan de Wenti shang bu Fayan ne (Why We Have No Voice on Chinese Overseas' Dignity Issue)," Bokee, Jan. 26, 2008, liangazy.bokee.com/viewdiary.22923404.html.

31. "Wo zai Ouzhou shuo Jiechu dao de Taiwan ren, Xianggang ren, Dongnanya Huaren (The Taiwanese, Hongkongese, and the Southeast Asian Chinese I Met in

Europe)," Tianya Community, Sep. 11, 2003, cache.tianya.cn/publicforum/Content/worldlook/1/2578.shtml.

32. "Huaren zai Nanyang (Chinese Overseas in the Southeast Asia)," Tianya Community, Feb. 16, 2008, cache.tianya.cn/publicforum/Content/worldlook/1/169017.shtml.

33. "Yuenan wei he yu Zhongguo Fanlian (Why Vietnam Turned Hostile Toward China)," Hexun Community, Feb. 27, 2008, ganzi13708557001com.blog.hexun.com/17246915_d.html.

34. "Jielu Sanshi nian lai Yuenan dui Hua Dazhanlue ji qi Zhiguo Lilian Dazhuanzhe (Revealing the Changes of Vietnamese China Strategy and Governance Ideas in the Past 30 Years)," Tiexue Forum, Feb 5, 2008, bbs.tiexue.net/post2_2571734_1.html.

35. "Jinqi Yuenan Dongxiang Zhide Zhongguo Gaodu Jiebei (China Should Watch out for Recent Movement of Vietnam)," Tiexue Forum, Feb. 13, 2008, bbs.tiexue.net/post_2559514_1.html.

36. "San yu yu lai, Women yin Ruhe Yingdui Yuenan de duoduo Tiaoxin (How We Deal With Vietnam's Provocation)," Bokee, Feb. 13, 2008, zgrdq.bokee.com/viewdiary.181107148.html.

37. "San yu yu lai, Women yin Ruhe Yingdui Yuenan de duoduo Tiaoxin."

9

Online Chinese Nationalism Toward the European Union

Economic and Diplomatic Implications of the Olympic Torch Relay Protests

Winnie King

INTRODUCING NEW PLAYERS, interests, and values into the policy-making process, globalization in the modern neoliberal international system has come to hold the state accountable to different values and interests upon multiple levels, including the domestic, regional, international, and transnational spheres. "Unlike many international relations theorists, we should not reduce international politics to interstate relations. Societies matter."[1] As illustrated by the events and subsequent responses following the European segment of the Olympic torch relay in April 2008, popular and supernationalism (or transnational nationalism) has the power to undermine the state's capacity to control and formulate national decisions. At the center of this have been the Internet and its multiple "online" channels of engagement, interest articulation, and mobilization. Including such mechanisms as blogs, messaging programs, bulletin board systems, chatrooms, and online forums, this chapter moves to examine the rise of online Chinese nationalism toward the European Union (EU).

Where "the core of the domestic-international connection is the impact of domestic institutions and interests on international interaction, and vice versa,"[2] debates of International Relations and International Political Economy have come to examine how domestic socioeconomic and political priorities, value systems, and institutions have come to dictate policy outcomes and state actions. Therefore, suggesting that an examination of this case of *online nationalism* necessitates an inherently international and transnational scope of analysis, it builds upon Michael Brittingham's view that "Chinese nationalism can best be understood as a reactive response to international events rather than domestic political manipulation."[3]

This chapter's examination of the Olympic torch relay, and its repeated disruption throughout Europe, delves into the rise of pro-Chinese nationalistic protests, anti-French boycotts across China, and consequent diplomatic exchanges between European and Chinese leaders. As a direct response to pro-Tibetan demonstrations, the overseas Chinese diaspora based in Europe drew upon the Internet and online resources to organize and challenge these activities. Led primarily by students, the Internet presented this group with a low-cost, high-profile means within which to articulate their concerns, and disseminate their views. For instance, in London "clashes between pro-Tibet protesters and police as the Olympic torch made its way through London"[4] led to the arrest of thirty-seven people. The events of April 7 in Paris, however, were even more extreme, with pro-Tibet independence protests and repeated and reasonably successful efforts to grab the torch itself, most notably from disabled torchbearer Jin Jing. This culminated in the forced extinguishing of the Olympic flame (three times), and led to the "humiliating" cancellation and shortening of the relay route itself.

Nationalism is not only reflective of state ideology, but also arises as a popular movement; as a protectionist response to negative events, or as a form of support. Therefore, it is not surprising these developments have elicited such a highly negative and nationalistic response. Perceived as more than mere criticisms against China, the overwhelming display of Chinese nationalism found on the Internet has been in direct response to what was seen as "anti-Chinese" reporting and imbalanced coverage of pro-Tibetan independence demonstrations by the European media, as well as calls from the European Parliament for world leaders to boycott the opening games of the Beijing Olympics.

The Internet and its multiple "online" channels for communication and information dissemination became a valuable forum in which netizens could share information across political and national boundaries. Furthermore, with the aim of defending China and condemning the Europeans[5] within two weeks this "virtual" dialogue had presented itself as a coordinated movement across Europe. What is more, and as detailed through this chapter, this rapidly spilled over from European protests to China's domestic arena in the form of boycotts, and ultimately drew diplomatic responses from European and Chinese authorities. While this case reaffirms what He calls "a sense of belonging to the nation, a sense of security, a feeling of national pride and attachment to the nation,"[6] it also sheds light on how this form of popular Chinese nationalism has had a destabilizing impact on Sino-European relations and foreign policy behavior.

Chapter Outline

Eliciting a string of diplomatic exchanges and dialogues to reaffirm commitment to the Sino-European relationship, this chapter sheds light on how international cooperation, interdependence, and China's participation in the global system has made it accountable to both domestic and international actors/institutions. Comprised of three sections, it begins with an examination of Chinese nationalism. Section 1 also moves to discern how Chinese diaspora nationalism has transcended nation and policy boundaries to produce coordinated European and transnational movements. It asks, how this movement has evolved from localized European protests into coordinated domestic Chinese boycotts of French products and firms?

In addition to discerning the rapid rise and resourceful organization of this movement, it also asks what these nationalistic movements and related activities have meant for EU-China diplomatic and foreign relations? And what, if anything, this could mean for China and its overall foreign policy strategy. Obtaining an understanding of Sino-European bilateral engagement is essential for appreciating the value and commitment European and Chinese officials have placed in this relationship. Section 2 therefore completes an examination of deepening relations up to the events of spring 2008.

It is important to note that the nature of this relationship is made more complex by the diverse character of the European Union itself. As an in-depth discussion of "What is Europe?" is beyond the scope of this chapter, "Europe" is defined by the context in which online mobilization activism arose in April 2008. Namely, while the primary boycott targets were French products and firms, protests were held in more than six European member-states. Furthermore, individual member-state and European officials alike have voiced concerns for the implications and costs of these activities. Lastly, as France held the position and authority of the European presidency during the April 2008 protests and demonstrations, its efforts to mend diplomatic ties were done in the name of both France and Europe.

The final section then examines how Sino-European diplomatic and economic relations have been affected by these mobilization efforts, as evidenced by concerns at the national, member-state, as well as firm level. Illustrating how online technology and information exchange mechanisms have successfully overshadowed diplomatic missions and meetings between European and Chinese officials, it assesses European and Chinese policymakers' responses within the context of international relations/foreign policy and economic relations.

Chinese Nationalism: State, Popular, and Supernationalism

Nationalism, by its very nature, is a term open in interpretation and necessitating contextualization. Whether perceived as a cultural, ethnic, or constructed concept, it follows that nationalism invokes responses of loyalty and protectionism from a government and a population. As Anthony Smith suggests, it "refers to the whole process of the growth of nations and nation states, sentiments of attachment to and pride in the nation, an ideology and language (or discourse) extolling the nation, and a movement with national aspirations and goals."[7] In the case of China, nationalism has been a central pillar in the leadership's efforts for nation building. Whether as a practice of reasserting legitimacy in its far reaching territories of Tibet and Xinjiang, distancing itself from past military and colonial humiliations, or reaffirming the "greatness" of China as a world and economic power, nationalism has played an important role in promoting and protecting national interests.

This line is embodied in traditional *state nationalism* in China. Driven by the party-state, it is within this ideological line that cultural symbols, ideology, and discourses of patriotism are identified. This framework of national goals and interests, developed by China's leaders, has therefore laid a foundation upon which *popular nationalism* has drawn its principles for mobilization. Within this context He and Gao's examination of Chinese nationalism, in particular *popular* and *supernationalism*, sheds light on their nature and how this form of mobilization may impose new restrictions of behavior on the state. The former comes "from below and represents unsystematic, popular national sentiments . . . [and] does not necessarily involve the support or active participation of the Chinese state, or may even involve the authorities' active efforts to stem the rise of nationalistic activities."[8] Similar to this, the latter is also rooted in the individual, but also underscores how "nationalism [also] consists, in part, of a sense of belonging to the nation, a sense of security, a feeling of national pride and attachment to the nation" which can transcend national borders and residency.[9]

Olympic Torch Relay Case Study: Protectionism, Patriotism, and Online Nationalism

With what He and Gao describe as *popular* and *super-* or *borderless nationalism*, the propagation, promotion, and protection of "China" and her interests are no longer limited to the state. As this case study therefore illustrates, nationalism can no longer be viewed as merely providing "legitimation for the Chinese nation-state, and stress[ing] individual loyalty to the state and the priority of the political community over individual rights."[10] For instance, international migration has improved access to information and expanded communication opportunities by placing participants beyond the jurisdiction of the state.

Offering netizens access to new and alternative channels of interest articulation, the Internet's resources pool has presented numerous advantages for a group as scattered and resource poor as the overseas Chinese students involved here. In addition to the traditional characteristics of information exchange and as opportunities for interest articulation offered by other political opportunity structures, "online" activities also provide anonymity, lowering the costs and risks of participation. Furthermore, the transnational character of the Internet as a forum for knowledge sharing places participants beyond state jurisdiction, and "under the radar" of detection, encouraging the entry of a greater number of activists. Most significant, however, is that this form of mobilization and organization has occurred independently of the state, and in the absence of state support.

The ease in which this form of engagement can take place can be evidenced by the level of coordination and the number of participants. Following Beijing's failed bid for the Olympics in 2000, the 2008 Games and related events had come to hold special significance of national pride in the eyes of the Chinese population. "Many netizens . . . [had said] that the Olympic torch represents the noble Olympic spirit; and the relay demonstrate[d] the ardent, global pursuit of excellence, friendship and respect."[11] Given the actions of anti-Olympic protesters and the negative reporting of the Olympic relay, Kenneth Lieberthal has suggested that for the Chinese, "regardless of how much China strives to become a constructive player in the world, 'many in the West will never accept that, [and] will seek to humiliate them.'"[12] Therefore, on April 19, almost two weeks following the London and Paris torch relay protests, demonstrators in London, Manchester,[13] Birmingham, Paris (three thousand participants), Vienna (fifteen hundred), and Berlin (three thousand),[14] followed by a demonstration in Belgium (one thousand) on April 29, came out in support of the Beijing games.

An estimated three thousand people attended the London rally, gathering in front of Whitehall and No. 10 Downing Street to protest against the media coverage of the April 6 Olympic torch relay and the Tibetan riots of March 2008. Organized and mobilized through the Internet participants, these "silent protests" were dominated by Chinese students reading for degrees in the United Kingdom and aimed specifically at the British Broadcasting Corporation (BBC).[15] In addition to banners that stated "BBC unfair," "Stop Media distortion, respect history, truth and China," and "Softkiller, don't spread hatred,"[16] the coordinators of the London protests also read out the following statement:

> Today we are here to oppose media distortion and media fabrication in the recent events. We are here in a quest for objectivity, fairness and justice. . . . As

> we heard, we saw, we say BBC . . . has misled the British public and the rest of the world by providing intensive untruthful reports and biased coverage . . . we are afraid that media distortion and media fabrication are going to build up the barrier and bring hatred between the Western world and the Chinese people.[17]

Wu Rui, a Chinese student and the coordinator of the Paris Rally in Support of the Olympics, stated that "during the relay of the Olympic flame in Paris . . . thousands of overseas Chinese were deeply offended and outraged by actions against the flame. . . . We are protesting against the stance of the European media, which did not provide fair and objective information."[18] Mobilized via the Internet, Wu pointed out that it provided the advantage of reaching a greater number of people, in different locations, as well as easing the process of coordinating simultaneous and similar events across Europe, most notably London and Berlin.[19]

Given the multiple interests and concerns of these participants, demonstrations of patriotism have come in several forms, including exhibited posters, placards, and chanted "Safeguard unification, oppose 'Tibet independence,'" "Tibet was, is and will be a part of China," and "Boycotting Olympics = New Berlin Wall."[20] Local petitions used to call for the correction of inaccurate stereotypes and misperceptions of both China and China-Tibetan relations were also completed online. They became a popular medium to articulate their support for the Olympics, with one online petition acquiring over 20 percent of its signatures from Europe.[21]

Revealing links between forms of nationalism, the supernationalism exhibited by pro-Chinese European-based protests quickly gave way to populist nationalism within China. Therefore, "notwithstanding the fact that the Internet reinforces surveillance and social control,"[22] blogs, chat forums, messaging, and text messaging, including such websites as Tianya.com, China .com, CE.cn, Xinhua Forum, Tudou.com, Sina.com, edu.QQ.com, bbs.cri .cn, and china.bbs.com, have been transformed into international and transnational sounding boards. It has given rise to "a significant platform for . . . venting anger with foreign attitudes toward China," as well as the posting of patriotic videos online detailing the crimes of the West against China.[23] Further searches through chat rooms, forums, bulletin boards, and petitions, in particular, revealed strong responses to the Olympic torch relay protests, including the use of such terms as "saboteurs."

As a result, attempts to express support for the Beijing Olympics, protest against Western media coverage, and promote the clarification of Chinese-Tibetan historical and political relations rapidly evolved into boycotts and petitions organized by way of the Internet. Within days and weeks of these initial pro-Chinese European-based protests, coordinated boycotts of French goods and firms came to dominate China's domestic Internet traffic (including Carrefour, Louis Vuitton, and Givenchy, and the cosmetics group L'Oreal). Representing a valuable resource pool for Chinese citizens to act, speak, and organize independently of the government, by providing those outside the "Great Firewall" access to information, it leads one to ask:

> If in fact an "anti-China" voice abroad could worsen China's international environment,[24] is it not also true that an "anti-West" voice abroad could severely undermine both China and its partner's capacity for engagement?

Evolving and Deepening EU-China Relations

Since the establishment of formal relations in 1975 with the European Community, China's relationship with the European Union has quickly evolved from limited trade relations into what has been labelled a "comprehensive partnership" in 1998, a "maturing partnership" in 2003, and eventually a "close partnership and growing responsibilities" in 2006. Reflecting the evolution in Europe's policy papers, this relationship has been founded upon several important factors, including complementary economic relations, cooperative and collaborative scientific and technological ventures/projects, and common strategic goals within the international system, as well as preference for multilateralism. These are embodied in the three pillars of Sino-European relations.[25] It was within this context therefore, that both Chinese and European leaders have demonstrated their commitment to this relationship.

Economic Relations

Economic relations have evolved into a multifaceted relationship, characterized by record trade and investment levels and a growing number of collaborative projects. Rooted in the complementarities of the European and Chinese economies, the former had been attracted by lower labor, resource, and production costs, as well as China's large domestic market. The Chinese were, and continue to be, interested in the contribution Europe's high-tech industries can make to the development of its economy and work force. In the area of cooperation and technological collaboration, economic relations between Europe and China have experienced an upsurge in foreign direct investment (FDI) levels and projects. Chinese FDI to Europe, beginning much later, has been dwarfed by the amount from Europe. For instance, by 2006 "China [had] approved 2,738 EU-funded projects . . . involving a contracted EU investment of 10.58 billion US dollars."[26]

In the area of trade, "trade volume between China and the EU [had risen] 25.3 percent to hit 272.3 billion US dollars in 2006."[27] As early as 2004 Europe and China had become the largest and second-largest trading partners respectively. For Europe however, although China had become "EU's fastest-growing export market, purchasing goods worth US$12 billion in 2007, up 36 percent from 2003,"[28] European authorities quickly began to raise concerns for its ever-burgeoning trade deficit. With trade growing at 20 percent annually since 2003, by 2007 China had a "surplus of $134.23 billion USD in the trade with EU . . . account[ing] for 51.2 percent of China's total trade surplus."[29] This, however, has not hindered the rise of cooperation and collaboration in the area of space (the Galileo satellite navigation program as well as the EU-China framework program, which promoted cooperation in the areas of science and technology) and aviation (with Airbus's decision to open an office in 1990, as well as the 2008 announcement of an assembly plant in Tianjin).

Multilateralism and Strategic Interests

Compared to the United States, Europe has been both positive and active in China's participation in the international arena. Particularly in the area of economics and trade, its support for China's admission into the World Trade Organization (WTO) reflected Europe's desire for gaining greater access to the Chinese market, while simultaneously promoting its own interests in international and multilateral institutions. Enthusiasm for this relationship is evident at both the member-state and European level. For instance, "reflecting the euphoric 'China fever' that is gripping Europe, the French government proclaimed 2004 the 'Year of China,' commemorating it with no fewer than 378 exhibitions and events around the country."[30] With the first EU-China summit held in London in 1998, EU-China level exchanges have also been steadily increasing. In fact, in 2004 "EU officials paid 206 visits to China . . . on average four times a week in order to exchange views with Chinese colleagues."[31]

For the Chinese leadership, participation within the international arena represents an opportunity to secure a place for itself as, first, the official representative government of China on the world stage—a direct nod to the "One China principle." Second, participation in international institutions provided means and opportunity for further social, economic, and policy development. Third, with its own interests in advancing the establishment of a multilateral or multipolar system, strong relations with Europe as an economic and political bloc plays a role in redressing imbalances in power with regard to the United States.

The mutual desire for normalized and effective communications and exchanges is also reflected in the establishment of the "Asia-Europe Foundation

(ASEF) in 1997 to promote mutual understanding and cooperation of the peoples of Asia and Europe through cultural, intellectual and people-to-people exchange programmes."[32] These forums for communication between Asian and European officials and institutions, such as during Asia-Europe Parliamentary Partnership Meetings (ASEP), as well as economic discussions during Asia Europe Meetings (ASEM), have also laid the foundation for developments at the bilateral level. For instance, the First China-EU Summit Meeting was held alongside ASEM 2 in London in 1998, a year which also saw visits from the seven heads of government of European countries including the United Kingdom, France, Italy, and President Jacques Santer of the European Commission to China.[33]

Further efforts to develop the bilateral Europe-China relationship are also illustrated by the institutionalization of this relationship in the form of numerous signed agreements. In the area of economics, efforts to renegotiate and upgrade the 1985 Trade and Economic Cooperation Agreement began in September 2006 in the form of the "Partnership and Cooperation Agreement" (PCA). The year 2004 saw the holding of "joint search and rescue exercise" between China's navy and French and British naval ships.[34] As section 3 details, however, these and other processes and missions have been overshadowed by the nationalistic activities, protests, and boycotts of April 2008.

Sociocultural Relations and Interests

This relationship has come to include more than just an exchange of goods and services. Sino-European relations are also characterized by extensive sociocultural exchanges, including growing levels of migration, tourism, and cultural and academic flows. In fact, with over 120,000 Chinese in Europe in 2006, there were "more Chinese students studying in Europe than any other destination in the world."[35] As one of the three central pillars of Europe's China strategy, the encouragement of people-to-people exchanges has come to reflect the value that communications and dialogue hold within the Sino-European relationship. For instance, in 2003 "the two sides signed a series of agreements, including one on group tourism."[36] Additional cooperative dialogues have also occurred in the areas of education, climate change, health care (in the instance of SARS and avian flu—H5N1), human rights, and the training of personnel in the area of law.

Implications for Sino-European Relations

The current Chinese leaders are actually facing a paradox. On the one hand, the government needs nationalism to boost its falling ideology legitimation based

> on Communism; on the other hand, it is very worried about the consequences of an overheated nationalistic sentiment among its people[37]

As the above quote clearly suggests, there is an innate contradiction between traditional Chinese state nationalism and the popular nationalistic activities and mobilization efforts of April/May 2008. No longer under the direct and sole influence of "government-sponsored patriotic education campaigns," the *international character* of the April protests therefore has had far-reaching implications for both the Chinese and European authorities.

In order to appreciate the motivations and rationale behind the European protests and demonstrations, as well as the Chinese boycotts, it is important to note the symbolic value and national significance of Tibet for China's territorial integrity, as well as the Olympic Games as a representation of China and the Chinese people's place in the international arena. With messages circulating on the Internet and mobile phones accusing companies of backing the Dalai Lama and supporting Tibetan independence activities, and given the large number of users with ready access to information, as well as the rapid pace of information dissemination, movements quickly grew into an economic and diplomatic predicament.

For Europe's trade relationship with China, the timing of these developments could not have come at a more inopportune moment. Following a November 2007 decision, "the European Commission sen[t] its largest-ever delegation to China in an effort to establish a new body to oversee trading relations—one designed to mirror the forum set up between Washington and Beijing."[38] This EU-China high-level mechanism is similar to the Chinese-U.S. Strategic Economic Dialogue. Beyond this bilateral dialogue, these online nationalism movements have also presented concerns at the domestic, regional, as well as interregional levels.

In addition to holding the Summer Olympics, 2008 was also a year of significant events and development for Sino-European relations and Europe-Asia relations, including the EU-China Summit in December 2008, and numerous ASEM events and related meetings, such as the Eighth ASEM Finance Ministers' Meeting (June 14–17), Fifth Asia-Europe Parliamentary Partnership Meeting (June 18–20), Eleventh Asia-Europe Business Forum (October 21–24), and Seventh ASEM Summit in Beijing (October 24–25). As a consequence, the Chinese government could not have afforded "to let the backlash continue too long without poisoning the atmosphere for the Beijing Olympics."[39] As mobilization, protests, and debates overtook the Internet within China's domestic arena, Chinese and European authorities, as well as business groups/interests, responded with multiple tactics to temper the implications of the torch relay debacle.

China's Response and Concerns

Recent popular nationalism and protests have not been a new development, as illustrated by the responses to the 1999 American bombing of the Chinese embassy in Belgrade and the 2001 collision between a U.S. surveillance aircraft and Chinese fighter jet. In fact, the utilization of the Internet and other technological tools were also evident in the rise of nationalistic mobilization efforts throughout the 2005 anti-Japanese protests following former Japanese Prime Minister Koizumi's visit to the Yasukuni Shrine. As Wasserstrom and others point out however, the pace and popularity of these most recent efforts have raised concerns with the Chinese leadership regarding how complaints against foreigners could potentially and quickly emerge, their capacity for being redirected toward domestic targets, and how they might lead to the "downfall of officials or entire regimes."[40] He goes on to point out that "China's leaders have found themselves confronted by . . . a desire to use popular nationalist sentiment for their own purposes (as a diversion from other kinds of discontent, such as worries about inflation) and a fear of losing control once crowds take to the streets."[41]

Scholars have, however, debated the value of the Internet and its capacity for enhancing the civic actor's capacity for interest articulation and/or challenging the government[42]: "Nationalistic sentiments, once agitated, need to be satisfied. This, in turn, puts great pressure on the government."[43] Minxin Pei, for instance, has suggested that "compared to before, the Internet has democratized opinion but . . . [that] fringe elements tend to exploit this new opportunity far more actively than the mainstream."[44] Acutely aware of the potential threat unchecked nationalism and mobilization would mean for state legitimacy, states are now faced with new and unforeseen restrictions on state foreign and diplomatic relations. "In fact, once the masses are moved by Nationalistic rhetoric, even a slight concession in international politics by the state could be interpreted as an act of treason and cause a legitimacy crisis for the regime."[45] In other words, "to allow popular nationalism to rise does not necessarily lead to the strengthening of the legitimacy of the regime. On the contrary popular nationalism can weaken the existing base of political legitimacy."[46]

These considerations have been reflected in the Chinese leadership's efforts to stem the rise and expansion of nationalism against Europe and European firms and products. Initially, state support for protests against the repeated disruption of the torch relay, or "journey of harmony," and Western media's efforts to tarnish the event by linking it to the Tibetan protests of March 2008, had been found throughout the national Chinese media. In fact, Chinese embassies in Europe had also lent their support to this organization of student movements. The Overseas Chinese Students for the Olympic Association, for instance, was provided the opportunity to mobilize and "form an unofficial guard for the Olympic torch when it passes through the UK."[47] In the United States, further support also reportedly came in the form of financial compensation, where students received as much as US$350 "from the Chinese government to protest against us (Tibetans and human rights activists)."[48]

As nationalistic activities quickly spread into unforeseen areas, such as economic and diplomatic relations, as well as taking on alternative forms of mobilization, however, the government moved to stamp out the nationalism-motivated activities. In this case, the focus was the Internet, which had been the primary mechanism for debating and disseminating information on the torch relay protests and media reporting: "On the Chinese side, the word 'Carrefour' has been censored on the web searcher Google, probably to protect Carrefour by limiting the access to pro-boycott forums."[49] Furthermore, in addition to blocking www.youtube.com following the Tibet riots in March 2008, the government has also censored domestic coverage of the Olympic torch relay. In fact, "some Chinese . . . reported difficulty sending text messages such as 'boycott Carrefour' and 'don't shop at Carrefour' on the cellular network run by China Mobile Ltd., the country's largest cell phone carrier."[50]

As calls went out over the Internet and phone networks for a boycott of French goods, with Carrefour as the primary target, it became apparent to the Ministry of Commerce that this form of consumer action could set a precedent for future incidents. This cannot be interpreted as the government's opposition to nationalism; rather, it had been acutely aware of the potential domestic repercussions, both with regard to economics and as a reflection of the state's capacity for controlling and maintaining order. With regard to the former, with tens of thousands Chinese employed at Carrefour and even more involved in the production of its products it also had potential repercussions for local manufacturers and workers. Ninety-five percent of the products sold by Carrefour in China were produced domestically, and with over one hundred stores in China it employed more than forty thousand Chinese;[51] the implications of these tactics were contrary to national goals of development and stability. Therefore, while the Chinese state had expressed an understanding for "a rational display of nationalism that does not run counter to China's trading interests,"[52] it quickly became concerned that "a strong, mass-based nationalism" would represent restrictions on its policy decisions and options, "distract its attention away from economic development, and, most dangerously, give rise to a civic consciousness that will spin out of control of the government."[53]

While the Chinese government suggested that "the informal boycott push ha[d] merit"[54] on April 16, by April 22 it indicated that "it would like to see the protesters tone down their demonstrations."[55] Therefore, while Foreign

Ministry spokeswoman Jiang Yu said the Carrefour demonstrations were "encouraging and touching," she added that "we do not agree with some people's radical actions."[56] Chinese state efforts to minimize the dissemination of the negative experiences of the Olympic torch relay, including censoring domestic media coverage and discouraging the spillover of boycotts to other industries and foreign firms, have been undermined by commentaries and efforts arising outside China. This potentially high cost could be observed in the case of General Electric (GE), where CEO Jeffrey Immelt pointed out that "the continuing global protests . . . [were] worrisome."[57] Although an American firm, the Chinese government's apparent inability to gain control over the potential spread of boycotts to include other French or foreign firms was a concern for international sponsors of the Olympics.

Therefore, claims that "the technological tools that were supposed to democratize China—websites, blogs, video sites and slickly produced television channels—are actually bolstering the Communist government by allowing it to mobilize anger at foreign critics"[58] are insufficient for explaining the events of April 2008. Blogs and chatrooms are no longer mere outlets for individuals to voice independent thought; rather, they have also become a setting in which to mobilize and organize.

The strength of nationalistic sentiments can be observed by the rapid rise and fall of Jin Jing, the disabled torchbearer harassed during the Paris torch relay. While initially portrayed as a national hero by both the public and the central government, her decision (though most likely at the suggestion of the Chinese leadership) to "oppose a mainland Chinese boycott of the French-owned Carrefour chain" quickly had "critics widely attack[ing] her: 'not only doesn't she have a leg, she doesn't even have a brain!'"[59] The government's attempt to regain a hold over the surge of nationalism, or manage these demonstrations, therefore was far from complete. With nationalistic movements reaching into regional and international politics and diplomatic relations, this case reflects how "external factors might influence a state's role, including the structure of the international system . . . and the rules, traditions, and expectations of states as expressed in the charters of international and regional organisations . . . multilateral and bilateral treaties; and less formal or implicit comments and understandings."[60]

Europe's Response and Concerns

In applying the term "Europe," particularly in the case of relations between Europe and China, it must be noted that "the officials of the Commission are more concerned with tangible results in economic cooperation and pay less attention to the differences of norms and values between them and their partners, whereas the European Parliament counts on the public opinion and is concerned about normative differences with its external partners."[61] Having called for a boycott in mid-March, in response to riots in Lhasa, the European Parliament had inadvertently provided a foundation upon which protesters could mobilize and challenge Sino-European relations. As a growing number of European and world leaders declared their nonattendance at the opening ceremony of the Beijing Olympics, however, it certainly did not instill popular Chinese confidence in the nature of Sino-European relations. These included UN Secretary Ban Ki-moon's decision, along with British Prime Minister Gordon Brown and German Chancellor Angela Merkel, *not* to attend the Olympic opening ceremony.

Perhaps, however, this "disunity" has provided the Commission with valuable space to utilize diplomatic mechanisms, and minimize the costs of these nationalistic efforts. Within two weeks of the protests and demonstrations across Europe, the week of April 21 saw the beginning of a number of intensive diplomatic exchanges between PRC, French, and EU officials and state representatives. On April 21, the first of a five-day high-level trip to China, the head of the French Senate Christian Poncelet met with Chinese leaders to both pass on President Sarkozy's apologies for the events and circumstances which took place on April 7 and reiterate France's stance on Tibet and the "One China" issue, as well as its support for the success of the Beijing Olympics.[62] In an effort to reduce anti-French sentiment and minimize the long-term political and economic fallout from these events, he read from a letter from President Sarkozy "what happened in Paris on April 7 has engendered a feeling of bitterness in your country. I want to assure you that the incidents that were brought about by a few people on this sad day don't reflect the feelings of my fellow countrymen for the Chinese people."[63]

This was followed by bilateral diplomatic exchanges between China and France and China and the European Union. Witnessing "a fresh round of shuttle diplomacy . . . the two countries have sent envoys to visit each other, conveying messages to respective leaders in the hope of mending the strained bilateral ties."[64] Not only did former French prime minister Jean-Pierre Raffarin, visiting for a forum on China-EU relations at the Chinese People's Institute of Foreign Affairs, stress Sarkosy's support for the One-China principle—"no matter the issues of Taiwan or Tibet, it is always a domestic issue for China"[65]—but he also emphasized Europe's desire for continued development of a partnership with China. Furthermore, in an effort to suppress "patriotic zeal [and] to concentrate on development,"[66] the Chinese also sent "senior diplomat Zhao Jinjun to France as a special envoy of Hu Jintao from April 18 to April 22, exchanging views with Sarkozy and other French senior officials on bilateral ties and issues of common concern."[67]

European Business

As stated in section 1, responding to international calls for a boycott of the Beijing Olympics, citizens both on and off the Chinese Internet called for their own boycott of French products, such as Carrefour, Louis Vuitton, and Givenchy, and the cosmetics group L'Oreal.[68] A "letter to all citizens" was circulated on the Internet, by e-mail and text message, calling for a May 1 boycott of Carrefour, stating:

> We suggest you go shopping in other supermarkets instead of Carrefour. If you have any inconvenience, we hope that at least you boycott Carrefour on May 1. The empty Carrefour on May 1 will send a message to the western countries that China should not be humiliated and the Chinese people should not be bullied.[69]

The decision to target France was based on events during the torch relay protests, incidents of protestors grabbing the torch, including the reasonably successful attempt against disabled torchbearer Jin Jing, as well as the "humiliating" results of the relay which resulted in the multiple extinguishing of the torch, the re-routing of the tour, and the cancellation of events and ceremonies. The ultimate decision was made after a four-hour long "chaotic journey through the streets of Paris, [when] Chinese officials finally called off the relay, with the torch reaching the Charléty Stadium" by bus.[70] In response, "the financial and economic channel of the popular portal www.netease.com posted a survey about a boycott of French products. Among the 43,880 respondents, 95.4 percent were in favor."[71]

In direct response to the rising boycott movement, and perhaps in an effort to call on other European officials for more responsible behavior, then European trade commissioner Peter Mandelson voiced political and economic concerns of linking Tibet protests to the Beijing Olympics, and the potential implications for "retaliation against European economic interests."[72] European and foreign businesses concerned with what informal "consumer sanctions" could represent for French investments and businesses (which have an estimated twenty billion euros worth of FDI in China) attempted to distance themselves from accusations of supporting Tibetan independence and a boycott of the Olympics. For instance, both the head of LV and Carrefour completed interviews with the French media denying any support for the Dalai Lama and their opposition to an Olympic boycott.[73] The boycott, however, quickly spilled over into other sectors and industries. For instance, as "the most popular European destination for Chinese tourists, with around 700,000 Chinese flocking to the country"[74] in 2007, France's tourist industry was also a victim of online Chinese nationalism. Travel agents in China went so far as to cut "French destinations out of their brochures, as Chinese anger continues to simmer," with the French Foreign Ministry eventually becoming involved.[75]

By the last week of April however, a growing number of European businessmen began to warn China's leadership that "a boycott of French products over pro-Tibetan protests abroad could spark a backlash against Chinese exports."[76] With calls for a boycott of French products "circulat[ing] mainly on the Internet . . . buoyed by the [PRC] government's initial blessing,"[77] the president of the EU Chamber of Commerce in China, Joerg Wuttke, highlighted the potential for retaliatory action on the part of Europeans against Chinese products. In fact, Katinka Barysch, of the Center for European Reform in London, has been quoted as saying, "The European trade commissioner has just warned of the tit-for-tat spiral—one side starts boycotting and then the other side starts boycotting and then before you even know it, a highly profitable trade relationship is disrupted."[78]

Conclusion

In the end, the long-term implications of these online protests and mobilization efforts were limited, coming to an abrupt halt with the Sichuan earthquake on May 12, 2008, which left eighty-eight thousand people dead or missing and five million homeless.[79] In examining this case, however, it has become apparent that "nationalism should not be treated as simply an epiphenomenon of domestic politics."[80] Where it has traditionally been a valuable mechanism for Chinese leadership efforts to promote national priorities, identity, and unity, as well as defining "China" as a nation state, nationalism today "[is] a particularly interesting force in China, given its potential not just for conferring legitimacy on the government but also for taking it away."[81]

This case of online Chinese nationalism toward the European Union has illustrated why nationalism can no longer be assessed within a limited scope of traditional Chinese state priorities and preferences. Rather, Dittmer and Gries suggest there is an increasingly populist character today, "less a political instrument of the Chinese Communist elite than 'old' nationalism."[82] Not only have participants gained access to new opportunities of engagement and interest articulation, but the Internet has also transformed the context for populist nationalism, offering access to transnational, international, and policy mechanisms. This has not only provided a solid foundation upon which to establish a legitimate and coordinated mobilization movement across national boundaries, but also for challenging both domestic and foreign authorities. In other words, the power and significance of popular nationalism is in its capacity to represent state priorities or undermine its long-term policy goals and international diplomatic and economic relations.

Government legitimacy, however, has remained rooted in support from the domestic arena and population, and maintaining a place within populist nationalism. Strictly "top-down 'state nationalism' theories fail to capture the dynamics of nationalist legitimacy played out between elites and the masses. In their efforts to maintain power, China's leaders have long paid close attention to popular nationalism."[83] As illustrated by this chapter's examination of both European and Chinese responses to the Olympic torch relay nationalistic movement, repeated high-level diplomatic and domestic populist sectors have been targeted to redress populist concerns and misunderstandings. For China and Europe, therefore, the deepening of economic relations and their evolution toward a more strategic partnership illustrates how "national disquiet intensifies in the age of globalization when national economies become interdependent"[84]

The rapid response and fervor of nationalism following the events of April 2008 calls attention to the fact that within the ideology of nationalism "the nation should be the highest object of political loyalty, should be politically independent (or at least autonomous), and whose interests must therefore be defended."[85] While Sino-European relations have evolved at an impressive pace, two of the three pillars of engagement have come under direct scrutiny. For Europe, concerns had been focused on the high cost for the economic and sociocultural arenas, as represented by the boycotts. China's efforts to regain control over popular movements however, hark back to its traditional tools of censorship and propaganda campaigns. This in itself reflects how Chinese nationalism has been transformed by its participation in the international arena; forcing Chinese leadership to balance its domestic legitimacy with its goals in the international arena.

Notes

1. Peter Hays Gries, "Tears of Rage: Chinese Nationalist Reactions to the Belgrade Embassy Bombing," *The China Journal*, no. 46 (July 2001): 28.
2. Jeffrey Frieden and Lisa L. Martin, "International Political Economy: The State of the Sub-discipline" (2001): 6, weber.ucsd.edu/~jlbroz/Courses/PS245/syllabus/frieden_martin_IPE.pdf.
3. Michael Alan Brittingham, "The 'Role' of Nationalism in China's Foreign Policy: A Reactive Model of Nationalism and Conflict," *Journal of Chinese Political Science* 12, no. 2 (2007): 147.

4. BBC News, April 6, 2008, news.bbc.co.uk/go/pr/fr/-/1/hi/uk/7332942.stm.

5. In this instance, anti-Chinese protests, biased media coverage, and demonstrations against the Beijing Olympics had occurred in several European countries and cities.

6. Baogang He and Yingjie Guo, *Nationalism, National Identity and Democratization in China* (Aldershot, UK: Ashgate Publishing, 2000), 5.

7. He and Guo, *Nationalism, National Identity and Democratization in China*, 1.

8. He and Guo, *Nationalism, National Identity and Democratization in China*, 2–3.

9. He and Guo, *Nationalism, National Identity and Democratization in China*, 5.

10. He and Guo, *Nationalism, National Identity and Democratization in China*, 1.

11. english.people.com.cn/90001/90776/90882/6389248.html.

12. Jayshree Bajoria, "Nationalism in China," *Washington Post* (online), April 28, 2008, www.washingtonpost.com/wp-dyn/content/article/2008/04/28/AR2008042801122.html.

13. www.youtube.com/watch?v=CQNeaZK3F_A&feature=related.

14. news.xinhuanet.com/english/2008-04/19/content_8011983.htm.

15. www.youtube.com/watch?v=BjcC6InqoKo&feature=related.

16. "Overseas Chinese in Britain rally in protest against Western media distortion," April 19, 2008, news.xinhuanet.com/english/2008-04/19/content_8011945.htm.

17. "Overseas Chinese in Britain rally in protest against Western media distortion."

18. english.cri.cn/2946/2008/04/18/1321@347757.htm.

19. english.cri.cn/2946/2008/04/18/1321@347757.htm.

20. news.xinhuanet.com/english/2008-04/19/content_8011983.htm; news.xinhuanet.com/english/2008-04/19/content_8012011.htm.

21. hi.news.sina.com.cn/news/xizang08/list.php?dpc=1.

22. Francoise Mengin, "New Information Technologies and the Reshaping of Power Relations: An Approach to Greater China's Political Economy," in *Cyber China: Reshaping National Identities in the Age of Information*, edited by Francoise Mengin (New York: Palgrave Macmillan, 2004), 7.

23. Geoffrey A. Fowler, Juliet Ye, and Loretta Chao, "China reins in nationalistic anger—Government stifles anti-West protests; Olympics play a role," *The Wall Street Journal*, April 24, 2008.

24. Yong-nian Zheng, *Discovering Chinese Nationalism in China* (Cambridge: Cambridge University Press, 1999), 88.

25. These are engaging China through multilateral institutions that enhance its participation in international affairs; intensifying bilateral Sino-European ties; and improving China's "domestic capacity" to govern. See David Shambaugh, "China and Europe: The Emerging Axis," *Current History* 243 (September 2004): 243–48; and Christopher Griffin and Raffaello Pantucci, "A Treacherous Triangle? China and the Transatlantic Alliance," *SAIS Review* 27, no. 1 (2007): 163.

26. www.chinadaily.com.cn/bizchina/2007-01/28/content_794691.htm.

27. www.chinadaily.com.cn/bizchina/2007-01/28/content_794691.htm.

28. ec.europa.eu/external_relations/china/docs/07_chronology_jan-07.pdf.

29. "Facts and figures about China-EU trade," April 28, 2008 english.peopledaily.com.cn/90001/ 90778/6400153.html.

30. Shambaugh, 245.
31. Benita Ferrero-Waldner, "Closing Remarks," in *The EU, China, and the Quest for a Multilateral World*, edited by Pierre DeFraigne (Beijing: China Institute of International Studies, 2006), 111.
32. ec.europa.eu/external_relations/asem/intro/.
33. www.chinese-embassy.org.uk/eng/wjzc/t27065.htm.
34. Shambaugh, 243.
35. www.chinadaily.com.cn/china/2007-10/24/content_6201018.htm.
36. Shambaugh, 245.
37. Dingxin Zhao, "Problems of Nationalism in Contemporary China: Student-Government Conflicts during Nationalistic Protests," in *Exploring Nationalisms of China: Themes and Conflicts*, edited by George C. X. Wei and Xiaoyuan Liu (London: Greenwood Press, 2002) 112.
38. Stephen Castle, "EU treads lightly on rights issue with China," *International Herald Tribune*, April 15, 2008, www.iht.com/articles/2008/04/15/business/trade.php.
39. Fowler, Ye, and Chao.
40. Jeffery N. Wasserstrom, "China's Inauspicious Year," *The Nation*, May 20, 2008, www.thenation.com/doc/20080602/wasserstrom.
41. Jeffery N. Wasserstrom, "Battle of the Beijing Boycotts," *The Nation*, April 28, 2008, www.thenation.com/doc/20080512/wasserstrom.
42. Gary D. Rawnsley, "Virtual China: The Internet as Threat or Opportunity?" *St. Antony's International Review* 3, no. 1 (May 2007): 42–57; Daniel Koldyk, "Virtual China: The Internet as Threat or Opportunity? A Response," *St. Antony's International Review* 3, no. 1 (May 2007): 58–63.
43. Zhao, 112–13.
44. Bajoria.
45. Zhao, 112–13.
46. Zheng, 91.
47. Caroline Davies, Ben Quinn, and Mark Townsend, "Pro-Chinese students to guard torch," *The Observer*, March 30, 2008, www.guardian.co.uk/world/2008/mar/30/olympicgames2008.china.
48. Phurbu Thinley, "China salaries overseas Chinese for anti-Tibetan protests," April 19, 2008, www.freerepublic.com/focus/f-news/2004254/posts.
49. euro-asie.sciences-po.fr/pdf_etudiants/newsletter/SilkroadsNewsletter_may2008.pdf.
50. Fowler, Ye, and Chao.
51. Christopher Bodeen, "European business officials warn China over consequences of boycott," *China Post*, April 23, 2008, www.chinapost.com.tw/news/2008/04/23/153296/European-business.htm. See also euro-asie.sciences-po.fr/pdf_etudiants/newsletter/SilkroadsNewsletter_may2008.pdf.
52. Tanja Vestergaard, "China, EU Inaugurate new Top-Level Political Dialogue Amid Tension over Olympic Protests," *Global Insight Daily Analysis*, April 25, 2008.
53. Zhao, 103.
54. "Europeans Weigh Likelihood of Chinese Trade Retaliation."

55. Bodeen.
56. Bodeen.
57. Evan Weiner, "Olympic Protests Could Have Economic Effect," *The New York Sun*, April 21, 2008, 18.
58. www.zonaeuropa.com/20080410_1.htm.
59. thechinabeat.blogspot.com/search/label/2008%20Tibet. See also Geremie R. Barmé, "Torching the Relay," *The China Beat: Blogging How the East is Read*, May 4, 2008, thechinabeat.blogspot.com/2008/05/torching-relay.html.
60. Brittingham, 150. See also K. J. Holsti, "National Role Conceptions in the Study of Foreign Policy," *International Studies Quarterly* 14 (September 1970): 246.
61. Jing Men, "EU-China Relations: From Engagement to Marriage?" *EU Diplomacy Papers* (July 2008): 17.
62. See www.france24.com/en/20080422-paris-honors-dalai-lama-chinese-dissident-France-china-protests&navi =France?q=node/1375990.
63. news.xinhuanet.com/english/2008-04/21/content_8020551.htm.
64. www.china-embassy.org/eng/gyzg/t429870.htm.
65. Liang Yan, "Former French PM: France wants to be China's best friend in Europe," *Xinhua*, April 25, 2008, news.xinhuanet.com/english/2008-04/25/content_8052299.htm.
66. Fowler, Ye, and Chao.
67. "China, France reach consensus on maintaining strategic ties," April 28, 2008, www.chinadaily.com.cn/china/2008-04/28/content_6646823.htm.
68. "Chinese Urged On Internet To Boycott French Goods," *Dow Jones International News*, 11 April 2008.
69. Aileen McCabe, "Chinese Internet awash with boycott tit-for-tat; French goods targeted as Olympic protests grow," *Montreal Gazette*, April 15, 2008, A1.
70. "Paris Protests Cut Short Olympic Torch Relay," April 8, 2008, www.france24.com/en/20080407-olmpics-torch-relay-paris-cut-short-protests-tibet-china-2008-games-france-beijing.
71. "Chinese netizens urge Carrefour boycott after torch relay incident," April 16, 2008, news.xinhuanet.com/english/2008-04/16/content_7989807.htm.
72. Castle.
73. www.youtube.com/watch?v=YtMIxJmTw1o.
74. www.france24.com/en/20080530-china-France-tourism-travel-agents-boycott-beijing-tibet.
75. www.france24.com/en/20080530-china-France-tourism-travel-agents-boycott-beijing-tibet.
76. Bodeen.
77. Bodeen.
78. "Europeans Weigh Likelihood of Chinese Trade Retaliation."
79. Edward Wong, "China: Hu Visits Quake Area Again," *The New York Times*, December 29, 2008, www.nytimes.com/2008/12/30/world/asia/30briefs-HUVISITSQUAK_BRF.html?_r=1&ref=asia.
80. Brittingham, 159.

81. Nicholas D. Kristof, "The World: Fruits of Democracy; Guess Who's a Chinese Nationalist Now?" *The New York Times*, April 22, 2001, query.nytimes.com/gst/fullpage.html?res=950CE7D71030F931A15757C0A9679C8B63&sec=&spon=&pagewanted=all.

82. Lowell Dittmer, "Book Review: China's New Nationalism: Pride, Politics, and Diplomacy by Peter Hays Gries," *Political Psychology* 26, no. 4 (August 2005): 652.

83. Gries, 28.

84. Toming Jun Liu, "Restless Chinese Nationalist Currents in the 1980s and the 1990s: A Comparative Reading of River Elegy and China Can Say No," in *Chinese Nationalism in Perspective: Historical and Recent Cases*, edited by George C. X Wei and Xiaoyuan Liu (London: Greenwood Press, 2001), 215.

85. Brittingham, 150.

10

Online Nationalism and Sino-UK Relations

Chun-wing Lee

THE OPINIONS EXPRESSED BY Internet users in mainland China have captured both media and scholarly attention in recent years. A number of publications have investigated the emergence of online nationalism and the differences among different expressions of nationalist sentiments by Chinese netizens.[1] These works show that the significance of online nationalism and other forms of public opinion that have emerged on the Internet should not be underestimated, since they could have an impact on the making of Chinese foreign policy.

Besides providing another case study on the nationalism being expressed by the Chinese online community, this chapter aims to fill an important gap in the existing literature studying online Chinese nationalism. This study examines the impact of online nationalism on the foreign policy of the People's Republic of China (PRC) by looking at its policy toward the United Kingdom (UK).[2] Concentrating on Sino-UK relations differentiates this study considerably from most other studies on online nationalism in China, since these are usually concerned with Chinese attitudes toward the United States and Japan. While it seems that, when dealing with the United States and Japan, the PRC does pay attention to the nationalistic sentiments found in the Chinese cyberspace, whether online nationalism can influence the PRC's policy toward the United Kingdom is another matter.

Moreover, the data source for this paper also differs from most existing studies on online Chinese nationalism, which usually study online political forums and online publications authored by scholars. Instead, this chapter analyzes the comments left by netizens on one of the most popular web portals in China: Sina.com. The major reason for choosing Sina.com is that, unlike many political websites or forums that often explicitly seek to encourage nationalist sentiments, Sina.com, as a web portal, may attract users from more diverse backgrounds. Whereas those who leave comments in response to news and articles concerning Sino-UK relations are probably more interested in politics than the general online community, netizens posting comments on Sina.com are more likely to represent viewpoints closer to the "general" opinion of the Chinese cyberspace community.

Sino-UK Relations

Before proceeding to our analysis, this section reviews the development of Sino-UK relations. The United Kingdom has played a significant role in China's modern history. The United Kingdom's victory in the Opium War in the early 1840s kickstarted the expansion of Western powers (and subsequently Japan) into China. During the late Qing period, the United Kingdom made war on China three times. In addition to defeating China in the First Opium War, thereby forcing the Qing government to cede Hong Kong Island to the United Kingdom, less than twenty years later British troops, joined by the French, again defeated China in the Second Opium War/Arrow War. During the Boxer Uprising in the late nineteenth century, the United Kingdom united with seven other nations to defeat China once again.

After the Second World War, the United Kingdom continued to rule Hong Kong despite the Chinese government's refusal to recognize the unequal treaties signed by the Qing government with the Western powers. Although during the Cold War, relationships between the PRC and the United Kingdom were unsurprisingly not close, from the perspective of the PRC the United Kingdom could be considered a lesser devil among the nations in the capitalist bloc since it was the first major power in the capitalist world to recognize the PRC. From the late 1970s onward, the future of Hong Kong became the major issue in Sino-UK relations, giving rise to a series of conflicts between the two countries. Even after the signing of the Sino-British Joint Declaration in 1984, which decided that Hong Kong would revert to Chinese rule in 1997, both countries clashed several times over their differences on the future of Hong Kong, the democratization process being a particularly contentious issue.

The first of July 1997 marked not only the return of Hong Kong to Chinese rule, but also the beginning of a new chapter in Sino-UK relations. As Hong Kong became a special administrative region of the PRC, what had been the major difference between Beijing and London was no longer an important issue.[3] In October 1998, then British prime minister Tony Blair visited Beijing and issued a joint statement with then Chinese Premier Zhu Rongji proclaiming that both countries would formulate an "enhanced, comprehensive partnership."[4]

One year later, Jiang Zemen became the first ever head of state of the PRC to visit the United Kingdom. During his visit, Jiang announced the principles for managing relations with the United Kingdom, which included cooperation in trade relations to enhance the interests of both countries and the development of interchanges in other respects.[5] From the United Kingdom's perspective, its policy toward the PRC had two objectives. The first was to help UK companies expand into the Chinese market, and the second was to promote political and social changes like democratization and the reduction of income inequality in China.[6] In other words, despite the inevitable differences between the two sides, both parties saw developing closer trade and business relationships as their first priority in their bilateral relations.

In July 2003, Tony Blair again paid a visit to China, this time joined by representatives from a number of major British corporations, including P&O, GlaxoSmithKline, BP, Shell, Rolls-Royce, Lloyd's of London, Barclays Capital, and Standard Chartered.[7] During his visit, Blair witnessed P&O sign a deal worth more than RMB 7.3 billion with the China Ocean Shipping Group Company (COSCO) and Maersk (a Danish conglomerate) concerning development of the port of Qingdao.[8] This visit can be regarded as another important step in Sino-UK relations. Soon afterward, the UK government created the China Task Force, which was chaired by the deputy prime minister. On the Chinese side, a Sino-UK Relations Task Force was also formed. The formation of these two task forces paved the way for further development of bilateral relations between the two countries. In 2004, PRC Premier Wen Jiabao visited London and another joint statement was released, with both sides agreeing to strengthen their relationship and become "comprehensive strategic" partners.[9]

With Sino-UK relations defined as a "comprehensive strategic partnership," relations between the two countries became closer and friendlier. Besides forming a comprehensive strategic partnership, the two governments agreed to arrange high-level visits annually and also to strengthen cooperation in various areas including science, technology, environmental protection, and education. The importance of this joint statement thus should not be underestimated since it signalled that, from the perspective of both governments, cooperation was preferable and both shared common interests.

Since the events introduced in the previous two paragraphs can be considered as the most important events related to Sino-UK relations in recent years, the following analysis will focus on how the Chinese online community expressed their nationalism during Blair's 2003 trip to China, as well as how they reacted to Wen Jiabao's 2004 tour to the United Kingdom, when he and Blair signed the joint statement defining the comprehensive strategic partnership between the two countries.

The 2003 Visit by Tony Blair[10]

During his July 2003 three-day visit to China, Blair visited Beijing, Shanghai, and Hong Kong. While in Beijing, he met with Wen Jiabao, Hu Jintao, and Jiang Zemin and answered questions from students at Tsinghua University.

This trip took place four months after the beginning of the U.S. war against Iraq. It was therefore not surprising to see that most of the comments posted online were related to the United Kingdom's role in the war. And Blair was largely perceived in a very negative light. Three related themes can be identified through reading the online comments. First, Tony Blair and the United Kingdom were disparaged for being a close ally of the United States. The following comments are examples:

> Blair is only a running dog owned by Bush.
>
> Actually, I don't like this Prime Minister. Because he is always following the US on all fronts, *doesn't act like* a man at all.
>
> In terms of personal ability and charisma, Blair is much stronger than Bush. But why is he willing to become a dog which always runs behind the US?
>
> He is not the British President [*sic*], but the Great Britain State Governor of the United States of America. The national leaders do not have to meet him.

Judging from these comments, we cannot understand Chinese netizens' perception of the United Kingdom without considering the United Kingdom's close relationship with the United States. Since in the eyes of the Chinese public the United States is often seen as a competitor or even an enemy of China, being a close ally of the United States would do the United Kingdom little good in gaining the affection of the Chinese people.

There appears to be another implication when Chinese netizens branded the United Kingdom or Blair as followers of the United States. As one comment voiced, "If there were no America, the whole world would be speaking German, [and] Great Britain would disappear soon." Another comment stated, "You think you are a major power because you are with the US, but actually the UK isn't as powerful as third-level countries. Only because of its relationship with the US, [they] still think they are in the nineteenth century." In other words, emphasizing that the United Kingdom is now a close ally or follower of the United States carries the implication that it is no longer the force it once was in world politics. Although I did not find any comments comparing the national trajectories of China and the United Kingdom, emphasizing the United Kingdom's decline could easily have the effect of boosting Chinese nationalism, since China, in stark contrast with the United Kingdom, is becoming a major player in world politics and the world economy after suffering more than a century of humiliation and decline.

But Britain's negative image, as perceived by Chinese netizens, does not derive solely from its friendly relationship with the United States. The United Kingdom's role in the Iraq War in 2003 undoubtedly made it less popular in China. Referring to Blair's meeting with students from Tsinghua University, a comment read, "What have we discussed with the invader?" Another comment said, "You devil killer should go to hell! The Iraqi people are crying!" Arriving in China just days after the death of David Kelly, the UN weapons inspector who was found dead days after giving evidence concerning the situation in Iraq to members of the British Parliament, the following comment was found: "[Blair] comes to China just to avoid the trouble [in his homeland] . . . On the issue of invading Iraq, he played an undignified role."

The Iraq War thus gave the United Kingdom and Blair the image of being an invader. In other words, besides being perceived as blindly following American foreign policy, and thus losing its former powerful status, the United Kingdom was disdained by Chinese netizens for its involvement in the Iraq War. Some more radical views even foresaw a future military conflict between China and the United Kingdom. One netizen remarked: "[Blair's] hands are full of blood. How can a killer sit in front of students. . . . Some day he will find a reason to attack any countries (including us)." An even more radical comment predicted a third world war:

> The US, the UK and Japan are an axis of the devil. These countries are rubbish, thugs and thieves. The Third World War is going to be a war between a "New Anti-Fascist Alliance" with China, Russia, France and Germany, etc., and a "New Fascist Axis" comprised mainly of the three aforementioned countries. People from the US and the UK shall rise and overthrow the Fascist regimes of both Bush and Blair. The final victory belongs to everyone in the world who loves peace.

Wars between China and the United Kingdom, of course, have happened before. The arrival of Blair just months after the start of the Iraq War therefore also prompted some Chinese netizens to remind their compatriots of previous conflicts between the two countries. Citing the defeat in the Opium War, the demolition of the Yuanmingyuan during the Second Opium War/Arrow War, and the presence of many Chinese treasures in the British Museum in London are three common ways for Chinese netizens to express their nationalism.

For example, one comment compared the Opium War with the Nanjing Massacre: "Remembering the Opium War is just like remembering the Nanjing Massacre! [They are] long-lasting hurts in Chinese hearts!!!" Responding to Blair's claim that starting the war on Iraq was the correct decision, a comment stated: "Does he also think that the Opium War was correct?" This statement attracted the following response: "To Blair, this hooligan and bastard, the Opium War was also correct and appropriate!!!"

The Yuanmingyuan was the Old Summer Palace of the Qing Dynasty, which was burned when British and French troops invaded Beijing during the Second Opium War in 1860. Under the PRC regime, the ruins of the Yuanmingyuan are now a patriotic education site. It is also widely seen as a symbol of the humiliation of the Chinese nation.

Since British troops were involved in the destruction of the Yuanmingyuan, many Chinese netizens mentioned the palace when commenting on Blair's visit to China. For instance, one netizen suggested, "I think the discussion between Blair and the Tsinghua [University] students should be held at Yuanmingyuan." Responding to another netizen who described Blair as "an English gentleman," one netizen commented: "English gentleman??? Ask him to visit the relics of the Yuanmingyuan!!! Tell him to leave all his British pounds before he departs!! [to make symbolic compensation]" The destruction of the Yuanmingyuan was also compared with the recent Iraq War: "The US and the UK invade Iraq, destroy the cultural objects and historical legacy [there], is this different from the invasion of China by the Eight-Nation Alliance, the fire which destroyed the Yuanmingyuan?"[11]

When the British and French troops destroyed the Yuanmingyuan, they also carted off many palace treasures, some of which are now in the British Museum. Losing its national treasures to the British was another symbol of China's national humiliation, and some Chinese netizens made such sentiments clear. One said, "The cultural objects in the British museum are the humiliation of our country." Another comment noted, "Tell [Blair] to return to China the cultural objects robbed by the British which can be found in the British Museum."

To sum up, during Blair's visit to China in 2003, Chinese netizens largely perceived Blair and the United Kingdom very negatively. Being a close ally of the United States, which had just started the war on Iraq, certainly did not help the British cause in the eyes of most Chinese. Britain was not only ridiculed for losing its world power status and thereby becoming the close ally (or follower) of the United States, but it was also strongly criticized for its actions in Iraq. Since the United Kingdom was seen as a close American ally, and because of its role in the Iraq War, some more radical sentiments even defined it as the enemy of the Chinese and predicted an eventual war between the two countries. Moreover, many Chinese netizens also used the humiliation suffered by the Chinese at the hands of the United Kingdom, including the Opium Wars, the destruction of the Yuanmingyuan and the holding of Chinese cultural objects by the British Museum, to demonstrate their discontent toward the United Kingdom.

But interestingly, while Chinese netizens were showing their contempt toward Tony Blair and the United Kingdom, hardly anyone was making concrete suggestions as to how the PRC government should deal with the United Kingdom. Almost no one was urging China to not sign business deals with the United Kingdom or take a more hardline approach toward it. In other words, the nationalist feelings expressed in response to Blair's visit to China in 2003 were largely sentimental and without any policy recommendations. The netizens, while showing their nationalism and anti-UK sentiments, were putting little pressure on national leaders to adopt any particular policy toward the United Kingdom.

Wen Jiabao's Visit in 2004[12]

In May 2004, Wen Jiabao visited five European countries—namely, Germany, Belgium, Italy, the United Kingdom, and the Republic of Ireland. Wen stayed in the United Kingdom for three days, his first official visit to that country since becoming premier. During his visit, a new Sino-British Joint Statement was signed declaring that the two countries would form a "comprehensive strategic partnership."

But before the signing of the statement, there had been virtually no indications that both countries were ready to become comprehensive strategic partners. Thus, on the Internet, or at least judging from the comments posted by those visiting Sina.com, few netizens discussed what Wen would achieve in the United Kingdom. Furthermore, since the official visit to the United Kingdom was only part of a larger European tour, the United Kingdom received no special attention from the netizens. The outpouring of nationalist sentiment appearing less than a year earlier when Blair visited China had almost completely disappeared. Although I found one comment that "[i]t would be great to bring back some cultural objects lost by China," the few comments that did urge Wen Jiaobao to achieve something in Europe were concerned with the arms embargo imposed on China by the European Union. In other words, Chinese netizens had few expectations concerning Wen's trip to Europe, particularly during his stay in the United Kingdom.

But after news was released that the statement had been signed and that the two countries had agreed to become comprehensive strategic partners, some anti-British sentiment became evident, although the level of emotion cannot be compared with the sentiment shown a year earlier during Blair's visit to China. While the negative image of the United Kingdom and Blair, as portrayed by many netizens less than a year earlier, did not completely disappear, it had largely given way to an image that perceived the United Kingdom as a country that China should not befriend owing to its unreliability.

One comment noted, "The UK is not reliable!!!!!!!!!! Be careful Premier!!!!!!!" Another comment described the United Kingdom as follows: "The British Empire is the most politically mature country in the world, most cunning!" Similar messages were given elsewhere; the following two comments recalled the close relationship between the United Kingdom and the United States:

> The UK is the best friend of the US, how can it be reliable?

> Can the British be trusted? They are the American dogs.

Some depicted the United Kingdom more negatively while questioning the decision to form a strategic partnership. One comment said, "When getting along with an old imperialist country like the UK, it's necessary to be more careful." This time around the netizens seldom mentioned the Iraq War, which was a hot topic in 2003, although the theme was by no means absent, as the following example shows: "'Strategic' the hell! Didn't the UK and the US attack Iraq without the endorsement of the United Nations? Will it form a strategic relationship with China? Wait, just a dream!!!"

A few comments did turn out to be a tacit criticism of the PRC's decision to establish such partnership: "China is collaborating with a tiger, China and Britain are two very different countries, should not establish a partnership, China has a Confucian thinking, which advocates peace and cosmopolitanism, and Britain is a robber country." Another comment also made its disapproval explicit: "Standing on the same side with a fox, doing dirty things together—what strategic partnership? Just tolerating its hegemonic actions, copying its human rights values." Probably because the posters expected other netizens to know that the United Kingdom had a bad track record in bullying China and other countries, neither comment tried to back up its negative image of the United Kingdom with any evidence.

Opinion in cyberspace, however, is not one-sided. Questioning the decision to form a comprehensive strategic partnership with the United Kingdom or criticizing the government cannot be considered the dominant view on the Internet. Many posters actually welcomed this event and therefore praised the authorities for what they perceived as an achievement. The following are several comments in response to the news of the signing of the new Sino-British declaration:

> This trip is worth it.
>
> Got a reward, worth it.
>
> Support the smart decision of the Central government, thank you Britain for its strategic consciousness.
>
> Although the strategic partnership between China and the UK does not have concrete content, it carries huge political meanings. Sometimes in international politics, the functions of political meanings are more important than concrete content. I support the country's current foreign affairs tactics.
>
> China is expanding its peaceful foreign affair tactics. Even though a strategic partnership is not everything, but using a form of common interest [*sic*] to impede the enemies, it's also a mature and intelligent foreign policy strategy employed by a major country.

Such comments showed the existence of an alternative viewpoint concerning the strategic partnership between the two countries. Whereas one set of opinions questioned the agreement or even opposed it, another set argued that such an agreement would be beneficial to China. Nationalism therefore does not necessarily imply the disparagement of other nations. For example, one comment regarded the reaching of such an agreement as a demonstration of China's rise:

> The strategic partnership between Britain and China is based on the context that China has continued its rise. People should know that there is nobody in this world who is willing to do a loss-making business. Also, there are no forever friends and forever enemies. There are only forever interests. This proves that the interests of China and Britain are getting closer.

Another comment implied that through this partnership, China could learn from the United Kingdom: "The Sino-UK relationship should be much more important than relations with other countries. For the sole reason that China would obtain substantial benefits, in terms of economic, social, political and scientific [aspects]."

Britain could even be portrayed positively after securing this partnership with the PRC. One poster praised the United Kingdom as follows and suggested that learning from it would be beneficial to China: "The UK is a world power, got advanced technology and an advanced economy. I believe we can learn a lot of things from them to develop the scientific standard and economic standard in our countries." Another poster also described the United Kingdom positively: "Hong Kong has been instrumental to China's opening and reforms, Britain returned a prosperous Hong Kong to China, Britain was the first Western power to have recognized the People's Republic of China, and has always respected the principle of One China."

Another reason to welcome this partnership was that securing the United Kingdom as a strategic partner would be a Chinese triumph over the United States. This viewpoint is understandable because, for many netizens in China, the interests of the PRC and the United States are antagonistic. Becoming a partner with the United Kingdom, a close ally of the perceived enemy of China, was therefore treated as a Chinese victory. One poster wrote, "Got an ally from the US." Another comment showed the same attitude: "Strategic partnership, the failure of the US!" In short, earlier we saw that in 2003, disparaging the United Kingdom as a fallen world power was a common way for Chinese netizens to express their anti-UK attitudes. But after a comprehensive strategic partnership had been reached between the PRC and the United Kingdom, such sentiment was hardly to be found. Instead, some netizens who supported this newly formed partnership acknowledged the United Kingdom's important status in world affairs, particularly because it was a close ally of the United States.

Nevertheless, even for these netizens who supported the agreement, not all thought that the United Kingdom was a country that deserved respect. For example, one comment said, "It's good to have got this thing [the comprehensive strategic partnership], at least on the surface. But must be aware of the nature of such an old imperialist empire, should not only look at its words, also got to look at its actions." Another commenter wrote, "As long as it is good for China, it is okay to co-operate with the United Kingdom, which is a hooligan country." The friendship between the United Kingdom and the United States was recalled by another poster: "Since the UK can be the running dog of the US, then the UK can also be the follower of China."

To sum up, two different opinions can be identified regarding the establishment of the strategic partnership between China and the United Kingdom. Some questioned this agreement, mainly because, from their perspective, the United Kingdom was not a trustworthy nation. On the other hand, others thought such a partnership would be beneficial to China or else was a demonstration of China's rise. Whereas those who questioned or opposed the partnership tended to view the United Kingdom negatively (although anti-UK sentiment was much less serious than that shown in 2003), those who supported the partnership perceived the United Kingdom either positively or negatively.

Worthy of note, Wen Jiabao visited not only the United Kingdom but four other European countries as well. Taking the comments posted by Chinese netizens in response to the trip as a whole instead of focusing solely on the United Kingdom, one would get the impression that the online community received this European trip well. Below are some comments that greeted Wen's return to China:

> Premier Wen! Greetings! Hard work! We understand all you have done for the motherland! The people thank you and love you.

> Very impressed by the . . . new leadership collective, visiting Europe and strengthening co-operation with Europe is a smart decision with long-term vision!

> Hard work, Premier! Be brave to make decision! Billions of people are supporting you!!!

Online Nationalism and Foreign Policy-Making

The reaction of Chinese netizens to Tony Blair's 2003 trip to China and Wen Jiabao's 2004 visit to the United Kingdom having been presented, this section analyzes whether netizen comments, particularly those expressing nationalist sentiments, had any impact on the PRC's policy toward the United Kingdom.

In 2003, when Blair visited China, the dominant view on the Internet was that he should not be welcomed because the United Kingdom, along with the United States and other allies, had just started the war against Iraq. The United Kingdom was also disparaged for having lost its world power status and being the "running dog" of the United States, and it was blamed for the sufferings of the Chinese people in the past. If PRC leaders had taken such sentiments into account when formulating policy toward the United Kingdom, it would have been difficult to understand why, less than a year later, the two countries became comprehensive strategic partners.

While it is tempting to conclude that the PRC leaders, at least on the issue of formulating policy toward the United Kingdom, were not interested in learning of the nationalist and anti-UK sentiments found on the Internet, a more careful survey of the online nationalism displayed in 2003 offers an alternative suggestion. As noted earlier, although there was an outpouring of nationalist sentiment when Blair visited China in 2003, these expressions were rarely attached to any policy suggestions. In other words, although the netizens made their attitudes toward Blair and the United Kingdom clear and therefore did not welcome his visit, they were not suggesting how their national leaders should deal with the United Kingdom in the future. The PRC leaders therefore faced little pressure not to continue to improve China's relationship with the United Kingdom.

When Wen Jiabao visited five European states including the United Kingdom in 2004, anti-UK sentiment was much less serious. This suggests that the nationalist feelings shown in 2003, which defined the United Kingdom as the significant other, emerged only because of the war against Iraq. When the war was no longer a hot topic, and when the main character in the news was Wen Jiabao instead of Tony Blair, the mood of the online community changed. Interestingly, it is also possible that Chinese public opinion would not fiercely oppose a closer relationship with the United Kingdom for two reasons. First, unlike its relationship with the United States and Japan, the relationship between the PRC and UK governments had not undergone any major crisis in recent years. Therefore, although the United Kingdom played a quite significant role in China's "humiliating" history and did not enjoy a good image in China because of its close relationship with the United States, the online community felt no need or urgency to demand action from Beijing when dealing with the United Kingdom. Second, the United Kingdom is a European country. Compared with the United States, Western powers from Europe do enjoy a more positive image among the Chinese people. When Wen visited the United Kingdom during a European tour, the United Kingdom was effectively presented as a member of Europe instead of as a close ally of the United States. The Chinese media, when reporting Wen's trip, named it a European five-nation trip, which had the effect of downplaying the United Kingdom's unique friendship with the United States and shifting public attention to the fact that Britain was a European country and a member of the European Union.

Such findings may also inform how both countries manage their bilateral relationships in the future. The Iraq War is no longer a dominant issue in international politics, and, judging from the improved image of the United Kingdom found on the Internet since its signing of the joint statement with the PRC, any future improvement in Sino-UK relations should not face much opposition from the Chinese online community. But if future events should lead to anti-UK sentiment reemerging in China, future important agreements between the two countries should be signed outside China to avoid public criticism from the Chinese online community; this way the attention of the online community will be focused on the role played by the Chinese leaders in the process and not on the British prime minister or the UK government. Indeed, such anti-British sentiment could be detected in early 2008, as the following section illustrates.

Reporting Tibet[13]

The previous sections showed that, since the Iraq War was no longer attracting widespread attention and both countries had signed the latest joint declaration, the United Kingdom's image in China had improved, and it could even be perceived positively by some Chinese netizens. But in early 2008, the United Kingdom's image was damaged from a Chinese perspective because of the way in which the BBC, along with other Western media such as CNN, reported the uprisings/riots in Tibet in March of that year. The alleged biased coverage by Western media angered Chinese netizens, and the United Kingdom was one of their targets. One major allegation against the BBC was that its website published a photo showing an ambulance and several medical personnel, but the English caption under the photo read, "There is a heavy military presence in Lhasa."[14] The following are some of the comments triggered by the perceived biased reporting found in the Western media:

> Western countries like Germany, the US, the UK, France will never be our friends, they hate the Chinese people fundamentally!

> Countries like the UK, the US, Germany and France have always been wolves, they are casting covetous eyes on us. We must be alert.

> Dalai is the organizer of the Tibet rebellion, imperialism [sic] like the US, the UK, Germany have not given up on breaking up China, these leading imperialists incite and support Dalai to cause troubles in China. Got to hit back in a strong way.

> The UK, France, Germany, Italy, the US, Japan, these Western powers, until today they still have not given up the ambition to destroy China!

Earlier we saw that the United Kingdom, when associated with being an American ally, was regarded with contempt by Chinese netizens. But when portrayed as a European country, the United Kingdom generally enjoyed a more positive image from the perspective of the Chinese online community. Here we see another image of the United Kingdom in the eyes of Chinese netizens: the United Kingdom is seen as a member of Western powers that are hostile to China. Recalling that modern Chinese history until 1945 was basically one of humiliation suffered at the hands of the West and Japan, such an image could easily appeal to many Chinese people when there was evidence that the West was still hostile to China, as shown by the perceived biased coverage of the Tibet incident by Western media, including the BBC.

Since the United Kingdom could be defined in such a way (although the proportion of hostile online comments explicitly mentioning the United Kingdom cannot be described as very high), its image had undoubtedly deteriorated quite a bit among Chinese netizens, as evidenced by the outcry targeting the West as hostile to China and supporting the independence of Tibet (which, from a Han nationalist point of view, means the breakup of China). For example, one netizen exhorted that "all compatriots should unite together to struggle against the anti-Chinese Western powers and the running dogs who are traitors!" Although not singled out for criticism, the United Kingdom should be classified as one of these "anti-Chinese Western powers" in such a context. In other words, the Chinese online community's anti-Western sentiment triggered by the alleged biased reporting of the Western media had the effect of making the United Kingdom less popular in China. Another implication of this episode is that, whenever any incident causes other Western powers like the United States and France to be viewed as hostile to China, the image of the United Kingdom, defined as a member of the Western powers, will also suffer. Therefore, although the main target of the online Chinese community appeared to be France during the subsequent Olympic torch relay in April 2008, it is highly likely that the United Kingdom's image in China would also "fall victim," not because of anything UK authorities had done, but because of what happened during the torch relay session in Paris and the French stance toward the issue of Tibet.[15]

Conclusion

By observing the online community's expression of nationalist feelings and Beijing's policies toward the United Kingdom, we may conclude that the importance of online public opinion on Beijing's foreign policy-making should not be overemphasized. The online community, while making their nationalist feelings evident, did not urge the government to take any specific action toward the United Kingdom while Blair was visiting China or Wen was visiting London. Thus, the lack of concrete policy recommendations on the Internet and the retreat of anti-UK sentiment in 2004 have given PRC leaders a free hand in handling Sino-UK relations without feeling pressure from online opinion.

Earlier I suggested that the lack of a crisis or serious conflict between the PRC and the United Kingdom, and the fact that the United Kingdom is a European country, may be two reasons that nationalist sentiments toward the United Kingdom are relatively moderate and rarely attached with policy recommendations. The implication is that because the United Kingdom is not perceived as the archrival of the Chinese, neither the general public nor the online community may be much concerned about long-term developments between Beijing and London.

Even though the image of the United Kingdom, along with that of the Western community, suffered a serious blow in China because of the controversies surrounding Tibet and the torch relay, we should not expect Chinese netizens to change their attitudes significantly about Sino-British relations because the controversies did not result from conflicts between the two governments. In fact, compared with other Western governments, the United Kingdom appeared to be friendlier to China, with Prime Minister Gordon Brown explicitly stating that the United Kingdom had no plans to boycott the Beijing Olympics.[16]

This lack of interest in Sino-UK relations of course does not mean that the online community is apathetic toward the PRC's foreign policy-making. Instead, while Beijing must manage relationships with many different countries in the world, it is reasonable to discover that Chinese netizens would pay more attention to relationships with certain countries like the United States and Japan, two countries that may be understood as public enemies number 1 and 2 from the perspective of the Chinese online community. In fact, one surprising finding when analyzing the online comments on Wen Jiabao's European tour in 2004 is that many directed their anger toward Japan without any reference to the relationship between China and Europe. This phenomenon prompted one poster to leave this following comment:

> This is [a place] for comments concerning "Wen Jiabao's visit to five European states and the EU Headquarters," but there are so many [people] scolding little Japan. Although it's off-topic, it's understandable. The Japanese are nothing, for a long time they have never stopped discriminating against the Chinese people . . . Defeat little Japan.[17]

Another significant finding of this chapter is that the image of the United Kingdom can be understood in several ways by Chinese netizens. The United Kingdom is simultaneously the first Western power to have humiliated China in modern history, a close ally of the United States, a European country, and a member of the European Union. When the United Kingdom's identity as one of the Western powers that have been (and some may claim still are) hostile to China is emphasized, the United Kingdom has a very negative image among the Chinese, as demonstrated during the Tibet episode. In addition, the comments about Blair's visit to China in 2003 also tell us that the Chinese online community will disdain the United Kingdom for being a U.S. ally. But if the Europeanness of the United Kingdom is stressed, it enjoys a much more favorable image in China. In other words, if the UK government wants to improve the United Kingdom's image among the Chinese people, it should be presented as a European country to the Chinese public.

I would like to end this chapter by pointing out several questions that future studies could address to further our understanding of online Chinese nationalism and Sino-UK relations. First, it is possible that political websites, which have a clear nationalist focus, may offer an alternative view on Sino-UK relations. A comparison between the content of nationalist feelings found on those websites and the more general opinion found on websites like Sina.com may help reveal different forms of nationalist sentiment in Chinese cyberspace. If such differences exist, their respective influences on foreign policy-making would be a very interesting research topic. Second, in April 2008, Chinese residing in the United Kingdom organized a number of demonstrations to protest the perceived biased coverage of the Western media, including the BBC, on the Tibet uprisings/riots and the Olympic torch relay. Future studies should take into account how these overseas Chinese, particularly the ever-increasing numbers of Chinese students studying in the United Kingdom, who appeared to form the majority of the protest participants, contribute to the development of Chinese attitudes toward the United Kingdom.

Third, this chapter has shown that the image of the United Kingdom in the Chinese online community can change depending on political developments. Future studies should therefore investigate whether the United Kingdom's identity of being a close American ally has become less significant in the eyes of Chinese netizens after becoming a comprehensive strategic partner of the PRC. In a similar vein, further studies should also examine the possible lasting impact of the Tibet episode on the United Kingdom's image in China. Last of all, existing studies on online Chinese nationalism have exclusively focused on the political dimension. The cultural elements of nationalism have so far been largely underexplored. In the case of Sino-UK relations, the English Premier League is closely followed by Chinese football fans. The news of England's defeat at the hands of Croatia in a European Championship qualifier in November 2007, which ended England's hope of reaching the finals, attracted more than seven thousand online comments on Sina.com.[18] The possible manifestation of nationalism and the image of England/the United Kingdom in such a context may deserve scholarly attention as well.

Notes

1. See Junhao Hong, "The Internet and China's Foreign Policy Making: The Impact of Online Public Opinions as a New Societal Force," in *China's Foreign Policy Making: Societal Force and Chinese American Policy*, ed. Yufan Hao and Lin Su (Aldershot, UK: Ashgate, 2005); Shih-Diing Liu, "China's Popular Nationalism on the Internet. Report on the 2005 Anti-Japan Network Struggles," *Inter-Asia Cultural Studies* 7

(2006): 144–55; Simon Shen, *Redefining Nationalism in Modern China: Sino-American Relations in the Emergence of Chinese Public Opinion in the 21st Century* (Basingstoke, UK: Palgrave Macmillian); Simon Shen, "'Holding Nationalist Flags against Red Flags'—Anti-American Icons in Contemporary China and their Reconstruction by the Public (1999–2003)," *East Asia* 24 (2007): 229–50; Simon Shen, "The Response of Chinese Intellectuals to the US War on Terror," *Journal of Chinese Political Science* 12 (2007): 237–80; Simon Shen and Mong Cheung, "Reshaping Nationalism: Chinese Intellectual Response towards Sino-American and Sino-Japanese Relations in the Twenty-first Century," *Pacific Review* 20 (2007): 478–97; Yongming Zhou, "Informed Nationalism: Military Websites in Chinese Cyberspace," *Journal of Contemporary China* 14 (2005): 543–62.

2. The Chinese usually call the United Kingdom *yingguo. Guo* means country, and the word *ying* was derived from "England." The Chinese terms for Britain—*buliedian*—or Great Britain—*Da buliedian*—are not commonly used. Even when referring to the British empire, the Chinese usually use the term *Da yingdiguo. Da* means Great; *diguo* means empire. Again, *ying* comes from England. In other words, the differences between England, Wales, Scotland, and Northern Ireland are rarely acknowledged apart from the sporting or football context, when the English national team would be called *yinggelan* instead of *yingguo.*

3. Shaun Breslin, "Beyond Diplomacy? UK Relations with China Since 1997," *British Journal of Political and International Relations* 6 (2004): 409–25.

4. The full version of the statement can be found at www.chinese-embassy.org.uk/eng/wjzc/zygx/t27071.htm.

5. "Bilateral Relations," Ministry of the Foreign Affairs of the People's Republic of China, www.mfa.gov.cn/chn/wjb/zzjg/xos/gjlb/1912/default.htm.

6. See Shaun Breslin, "Beyond Diplomacy"; Foreign Affairs Committee, House of Commons, *East Asia: Seventh Report of Session 2005–06* (London: The Stationary Office Limited, 2006).

7. Mary Hennock, "Blair's Business Relief in China," BBC, news.bbc.co.uk/2/hi/business/3072729.stm.

8. *Wen Wei Pao*, 22 July 2003: B8.

9. The full version of the statement can be found at www.chinese-embassy.org.uk/eng/wjzc/zygx/t101947.htm.

10. Online comments discussed in this section can be found at comment4.news.sina.com.cn/comment/skin/default.html?channel=gn&newsid=1-1-1378837&style=0, comment4.news.sina.com.cn/comment/skin/default.html?channel=gj&newsid=1-1-1391827&style=0# and comment4.news.sina.com.cn/comment/skin/default.html?channel=gn&newsid=1-1-1384654&style=1.

11. The "Eight-Nation Alliance" refers to the foreign troops that defeated China in the Boxer Uprising.

12. Online comments discussed in this section can be found at comment4.news.sina.com.cn/comment/skin/default.html?channel=gn&newsid=1-1-3196683&style=0, comment4.news.sina.com.cn/comment/skin/default.html?channel=gn&newsid=1-1-3191822&style=0, comment4.news.sina.com.cn/comment/skin/default.html?channel=gn&newsid=1-1-3197479&style=0 and comment4.news.sina.com.cn/comment/skin/default.html?channel=gn&newsid=1-1-3181560&style=1&list=1#.

13. All online comments discussed in this section can be found at comment4.news.sina.com.cn/comment/skin/default.html?channel=gn&newsid=1-1-15198897&style=0# and comment4.news.sina.com.cn/comment/skin/default.html?channel=ty&newsid=6-12-3310086#.

14. See news.sina.com.cn/c/p/2008-03-21/232315198897.shtml.

15. The torch was extinguished in Paris because of the protests against the PRC's rule in Tibet. French President Sarkozy threatened to boycott the opening ceremony of the Beijing Games and the mayor of Paris, Bertrand Delanoe, announced that he wanted to award the Dalai Lama an honorary citizenship of the French capital.

16. I even found online comments made by Chinese netizens praising the effort of UK police to arrest pro-Tibet protestors who had tried to disrupt the torch rally in London. For example, one netizen wrote, "The UK Police have done a great job, Great China thanks you." Another message was like this: "Thank you, British Police, must not take a soft stance against these people who support the independence of Tibet, must arrest these pro-independence rubbish." Both comments were found at comment2.news.sohu.com/viewcomments.action?id=256123615&pageNumber=3.

17. This comment can be found at comment4.news.sina.com.cn/comment/skin/default.html?channel=gn&newsid=1-1-3181560&style=1&list=1#.

18. See comment4.news.sina.com.cn/comment/comment4.html?channel=ty&newsid=6-12-3310086.

11

A Constructed (Un)Reality on China's Reentry into Africa

*The Chinese Online Community Perception of Africa**

Simon Shen

SINCE THE INTERNET, as we now know it, officially started in 1994, its use in China has developed at an exponential rate. At the end of 2007, there were approximately 210 million online citizens in China, only slightly fewer than in the United States.[1] In a country where full democracy and complete freedom of expression are lacking, online communication arguably plays a more significant role than it would in countries like the United States. When it comes to topics that are not frequently addressed in the authoritarian public domain of China, the flow of information within the online community, even though it is frequently biased and surreal, is paramount in shaping perceptions among the ever-growing number of Internet users.

To Beijing's way of thinking, China has never left Africa. However, the general perception is that after the end of the Cultural Revolution in China and with the reversal of a Chinese foreign policy that had heavily patronized African countries under the slogan "Third World Unity" during the Maoist era, Sino-African relations were given little attention in the 1980s and 1990s. Following China's much-heralded twenty-first-century reengagement and reentry into Africa, for reasons greatly different from those of the 1960s, the Chinese perception of Africa, as expressed by the online community, exemplifies this lack of relationship continuity. In the academic domain of African studies, the opinions of Chinese Internet users have been little investigated; in the domain of Internet politics, the reconstructed image of Africa by Chinese Internet users and its inherent misperceptions have also been neglected.

This chapter is the result of systematic qualitative research on the online community in China, and aims to bridge the gap between formal studies of

Sino-African relations and their perception online by reconstructing the online image of Africa as created by Chinese Internet users. Such a reconstruction is instrumental in understanding not only contemporary Sino-African relations from the popular perspective, but also the distorted nature of information circulating in Chinese cyberspace. This chapter is in three sections. The first reviews relevant literature on past and contemporary Sino-African relations, details the studied images of Africa in China, and examines the involvement of the online Chinese community in Chinese diplomacy. The second analyzes the different aspects of Africa as modern Chinese Internet users imagine them to be, represented by how they stereotype the Africans as inferior Chinese partners, disciples of the Chinese, low-priority financial recipients, and automatic supporters of China who are generally "poor," "lazy," "sexist," and "threatening." The final section examines the implications for Sino-African relations and the online Chinese community.

Primary Sources to be Studied

The time frame of the article covers the period from November 2006, when the third Ministerial Conference of the Forum on China-Africa Cooperation (FOCAC) was held in Beijing and when the first heads of state summit was also held, until August 2008, when African leaders met Chinese President Hu Jintao after the Beijing Olympics opening ceremony, and before the Obama fad gained momentum in China in October 2008. Within this period, the chapter identifies below the Internet discussion forums in China that were used to gather empirical data.[2] Online discussions over the almost two-year period on the chosen topic, which amounted to approximately one thousand directly relevant threads, have been carefully studied.

If the topic of study were a geographical area more familiar to Chinese online users, like the United States or Japan, locating the right online forums as primary sources would be relatively easy. However, when the topic is Sino-African relations, relevant discussion appears in different forums only in a far more scattered and spontaneous manner. From such constraints, the forums that we have chosen primarily include:

- The *Qiangguo luntan* (Strong Nation Forum, SNF), arguably the most famous meeting point of Chinese nationalists since the Kosovo War in 1999 [bbs1.people.com.cn/boardList.do?action=postList&boardId=1];
- The *Feizhou Buluo Wang* (African Blog, AB), a single-focused forum on issues relating to Africa [www.africaren.com];
- The *Tianya Shequ* (Tianya Community, TYC), one of the most visited mainland Chinese websites by Chinese in Hong Kong and overseas [www.tianya.cn];

- The *Wangyi Luntan 163* (Easyweb Forum 163), a combo website featuring all sorts of information for general readers [bbs.163.com];
- The *Maimop Da Zhihui* (Collection of Cats' Wrestling, DZH), a leisure-oriented social issue forum targeted at teenagers [dzh.mop.com];
- The *Liuliu Wang* (Go Go Web, GGW), a forum on international travel [www.guolv66.com];
- The *Tiexue Shequ* (Community of Iron and Blood, CIB), a forum focusing on military affairs [bbs.tiexue.net]; and
- YouTube, the popular worldwide web 2.0 audiovisual interactive website; this website has also been consulted in the hope of understanding how the Chinese select visual presentations on Africa online (although actual materials on the topic on YouTube are severely limited) [www.youtube.com].

The above websites do not cover all online Chinese discussion on the perceptions of Africa. Yet, collectively, given their popularity among Chinese, they nonetheless form a comprehensive macro picture. For instance, SNF and TYC are the two major websites that are visited daily by many young Chinese Internet users. With registered members numbering more than 570,000 (as of November 2007 for SNF) and 20,000,000 (as of 30 June 2007 for TYC), and a daily hit rate of 1,500,000, a small thread on Africa on these forums may create a great perceptive impact.[3] During our research, we found that the perceived image of Africa in the more specifically Africa-focused websites like AB was very similar to that expressed in the generic websites.

Before looking at these sources, we must first acknowledge the methodological limitations of the chapter. The Internet aims at promoting a free flow of information across geographical boundaries and beyond known identities. Yet we can only assume those using simplified Chinese characters (coded in "GB2312") in the above forums to be Chinese cybermen, although in reality Internet users of other nationalities may also post messages there. Moreover, threads in the forums—TYC in particular—sometimes contain fictionalized messages in a half-novelized manner when they touch on Africa, making the credibility of the content difficult to verify.[4] Nonetheless, these messages are still valuable in understanding the perception of the Chinese of Africa, even though some of the stories are likely to have been fabricated.

Sino-African Relations and the Online Chinese Community: Literature Review of the "Missing Link"

Many researchers have studied the online Chinese community and its relationship with domestic politics and democratization. Most take the line of Geoffrey Taubman, who predicts that the Internet will challenge the ideational and organizational control of China's party-state.[5] Sometimes, this challenge can be indirect and subtle, as dissidents may use the strategy of *dachabianqiu* (kicking the ball from the side) to criticize the party-state by using remote topics.[6] Characterizing the Internet as serving Habermas's conception of a "public sphere," Yang Guobin argues that the relatively free chat rooms, listservs, newsgroups, electronic magazines, and bulletin boards are already forming such an arena in China.[7] Most analyses of the online community in China focus on its impact on information flow and the growth of Chinese civil society, or even the popularization of dissenting ideas, such as Falun Gong.[8]

Online discussion of Chinese foreign policy is believed to have had an impact on the topics of discussion in the following ways. First, as Peter Gries and Shih-Diing Liu respectively argue, when a series of anti-Japanese demonstrations broke out in China in April 2005, online mobilization was the major catalyst.[9] They claim that this online mode of mobilization bypasses the censorship of the party-state in China, and represents a bottom-up approach of popular nationalism (*renmin minzuzhuyi*).[10] Second, online discussions may have specifically influenced the policy-making elites in China, who may adopt ideas first aired by Internet users to formulate Chinese foreign policy. For example, Chinese scholar Zhou Me argues that the sudden cancellation of her scheduled meeting with the Japanese Prime Minister Koizumi, as well as the schedules of other high-level meetings, can be viewed as being influenced by Chinese online protests about China's foreign policy.[11] Third, some radical users may hack into what they see as anti-Chinese websites, involving "collectivist tendencies and links to state and corporate establishments," and pose a potential danger for Chinese foreign relations.[12] However, how such discussions of foreign affairs reinforce the stereotypical image China has of the outside world remains an issue to be studied.[13] For seemingly remote regions such as Africa, studies on how online discussions among Chinese users influence their perceptions of the African continent are at best given scant attention and at worst are almost nonexistent.

A New Phase in Sino-African Relations for the Twenty-First Century: Reentry or Recharge? Academic Explanations in the Non-Online Community

Contemporary Sino-African relations date back to 1955, when then premier Zhou Enlai of the People's Republic of China attended the Bandung Conference, followed by the establishment of formal diplomatic relations between China and Egypt in 1956. When the PRC was admitted into the United Nations in 1971 at the expense of the Republic of China (Taiwan), African votes were crucial. During the Cultural Revolution, China's support of her African comrades against imperialism (best symbolized by the huge investment in, but poor outcome of, the Tanzam Railway [*Tazara*]) was a cornerstone of Chinese diplomacy.[14] Sino-African relations started to wane after 1976, as most of the political and economic investments made by China in Africa proved to be of doubtful value. Fighting for formal diplomatic recognition from the Africans for the purpose of competing with Taipei's diplomacy also became increasingly unnecessary. However, after China's economic reform produced its early successes in the late 1990s, rapprochement between China and Africa was achieved. May 1996 can be identified as the milestone marking China's reentry into Africa, when President Jiang Zemin visited Africa and presented a five-point proposal for Sino-African relations—namely, "reliable friendship, sovereign equality, nonintervention, mutually beneficial development, and international cooperation."[15] This position, as Alden argues, is a significant move away from Mao's legacy, replacing the ideological motor with the concept of mutual benefit.[16] What, however, are the reasons that lie behind Beijing's new African campaign? Although the explanations offered are interrelated, rather than coming from totally opposed schools of thought, the existence of nonmonolithic views and interest groups on Africa within and beyond the Chinese borders has meant that different scholars have given different emphases.

Of all the explanations, two prime rationales repeatedly appear: those of energy and market. Scholars who support the resource maximization assumption, such as Erica Downs and Ian Taylor, argue to different degrees that the campaign primarily results from considerations for the supply of natural resources.[17] As one of the fastest growing economies in the world, and the world's second-largest consumer of oil, China is attracted to Africa's abundant resources to help solve its own economic problems; it has been a net importer of petroleum in general since 1993. According to a projection by the International Energy Agency, China's net oil imports will jump to 13.1 million barrels a day by 2030.[18] Chinese oil imports from African countries have increased from one-fourth of the total in 2004 to one-third in 2006. Recognizing the urgency of these issues, China's Ministry of Foreign Trade and Economic Cooperation (MOFTEC) has twice explicitly encouraged Chinese investors to go to Africa, in 2003 and 2005, in the hope of securing a long-term supply of resources.[19] In November 2003, Chinese President Hu Jintao declared that oil and finance constituted the two major components of China's national economic security.[20] China used to rely heavily on the Middle East for oil, which provided 40–50 percent of China's total oil imports, but over 75 percent of this went through the Malacca Strait, a sea route which is vulnerable to hostile action by external powers and terrorist attacks.[21] As David H. Shinn argues, the contribution from the African continent to China will continue to increase, as "China is reluctant to become excessively tied to the Middle East as a source of oil.[22] China also lacks refinery capacity for the heavier crude that comes from the Middle East." Yet we should acknowledge the fact that, as noted officially, the concurrent flames of a "China threat" and a conspiracy of "plundering" African natural resources have been discussed by many Africans, especially of the younger generation.[23] Anti-Chinese feeling has already led to protests and scattered violence in Africa—for example, in the Zambian copper belt and Lesotho. Partly as a result, kidnapping of Chinese is more frequently reported in recent years.

While acknowledging the role of energy, other scholars such as Andrea Goldstein and Domingos Jardo Muekalia give equal weight to the lure of the potential markets and investment opportunities in Africa to explain contemporary Chinese interest in the continent.[24] According to Goldstein and Muekalia, Sino-African trade could make a significant impact on the Chinese domestic economy. For instance, low-cost Chinese merchandise could be sold through the growing network of trading posts across urban and rural Africa. Through direct investment in Africa-based firms, the Chinese could take advantage of the mutual agreements between Africa and the United States (like the U.S. African Growth and Opportunity Act, AGOA) and between Africa and the European Union (like the Contonou Agreement), in order to export textile and agro-industrial products to other parts of the world via Africa.[25] They argue that this is why both the Beijing government and Chinese companies of different sizes have an interest in developing Sino-African relations, even though the general public may not share the same vision.

At the same time, forging strategic partnerships with the Africans is seen by many Chinese scholars, such as Jianbo Luo and Yongping Zhang, as Beijing's means of launching, or responding to, a diplomatic battle with other status quo powers, especially the United States.[26] Besides the competition in economic and geopolitical terms, the African votes in multilateral forums represent scarce goods that Beijing is keen to obtain. Indeed, China and most African states share similar viewpoints on issues like human rights, democracy, and sovereignty. Gaining African support on such issues might effectively put China into a leadership role of the Third World.

On the less tangible but equally important identity-building front, we should not neglect the fact that symbolic diplomacy and developmental assistance are still part of the key issues behind China's interests in Africa. Scholars such as Barry Sautman and Hairong Yan point out that the contemporary Chinese position originates from its identity as a developing, socialist, and former semi-colonial country.[27] Compared with other foreign interests in

Africa, China is, quite often, believed to be thought of by the Africans as the "lesser evil," allowing Chinese leaders to claim the Africans as "fair-weather friends."[28] The direct funding from Beijing to African civil services or the Chinese debt-waiving agreements with the Africans all serve such public relations purposes. These measures are meant to project a special message for the Africans. Unlike the assistance given from Washington, that from Beijing does not come with attached reforms aimed at changing tradition in Africa, and thus should be seen by Africans as a more welcomed alternative. Whether this suggestion is overwhelmingly accepted by contemporary Africans or not, which is not very likely, it is indeed the rationale behind Beijing's recent "peaceful development" (*heping fazhan*) diplomatic slogan—the latter replaced the more zero-sum implied "peaceful rise" (*heping jueqi*). The slogan was coined in the hope of claiming more credits for China as a responsible state in the global arena, without violating its principle of noninterference in the domestic politics of other sovereign states.

Probably all these rationales were in some way behind Beijing's calculation to reenter Africa in the late 1990s. The Chinese government has not denied any of these reasons for its African interest. As Andreas Lorenz reminds us, PRC leaders always work as traditional rationalists, pursuing only eternal interests but not eternal friends.[29] But subtle questions remain. How do ordinary Chinese, as expressed by the online community, view China's reentry into Africa? Do their perceptions fit any of these academic assumptions, or do they come up with other justifications? Might their perceived images of Africa affect Sino-African relations? These are the questions to be addressed by this chapter.

Constructivism and Internet Response

Viewed as an "off the map" continent for most Chinese, Africa has undeniably failed to draw particularly fervent attention from Internet users. Most Chinese know little about African history and politics besides what is portrayed in popular culture. High-level political exchanges between China and Africa, such as the ministerial conference of the FOCAC, have gained little attention online. Even the recent kidnapping, and deaths, of Chinese workers in Ethiopia and Nigeria aroused relatively little controversy online, especially when compared with the heated Internet response toward every tiny move of the Japanese rightists. Nonetheless, online discussions on the topic may be found with prudent research. What then are the key differences between the Internet opinions on China's reentry into Africa, and the academic explanations cited above? Putting aside the energy, market, partnership, or friendship issues, when commenting on China's reentry into

Africa, it was largely the constructivist line that caught the nationalist fancy of Chinese Internet users.

Briefly, constructivism emphasizes the role of constructed images and self-image imposed on other nations, rather than absolute interests. According to leading constructivist Alexander Wendt, "anarchy is what states make of it," because "there is no 'logic' of anarchy apart from the practices that create and instantiate one structure of identities and interests rather than another; structure has no existence or causal powers apart from process."[30] Realist interpretations of national interest-maximization in the international order did exist to a certain extent, but as explained by constructivist Douglas Porpora, they were only useful in explaining international relations when they were "embedded" into the "social knowledge" of a country.[31] Politicians, as well as fellow Internet users, could well apply constructivism to explain the mechanism of international relations as a strategy: a strategy designed to construct an image of reality. When a nation wishes to advance its national interests by constructing a desired national image to be taken up by the international system, it may intentionally use constructivist techniques—such as reconstructing reality—to create a more appealing national image, which, however, often deviates from the public perception found in the same nation. When China constructs itself to become a responsible state that is allegedly peacefully rising or peacefully developing, whether the Chinese in reality present such an image is quite another issue. The online Chinese perception on Africa and the Africans can be regarded as a typical example of such, which will be discussed as follows.

The Dominance of Constructivist Arguments: Analyzing Online Responses Toward Sino-African Relations

Based on the history of Africans living in China, as well as the Chinese experience in Africa, the traditional Chinese stereotyping of and discrimination against Africans in general is best studied from the works of scholars like Philip Snow, Frank Dikötter, Emmanaul Hevy, Barry Sautman, and Adams Bodomo. In his classic book *The Star Raft*, ideas in which are echoed by some of the aforementioned scholars, Snow gives readers a three-stage historic development of the image of Africa in China up to the end of the Cold War.[32] Since Snow's depiction is repeatedly referenced in our online study's constructed image of Africans, his work deserves to be discussed first and foremost here.

According to Snow, in the first stage of Sino-African relations in precolonial times, the dark-skinned Africans were known to the Chinese by the name of *Kunlun*.[33] By the ninth century, the word *Kunlunnu* (Kunlun slave) was attached to the Africans who were brought by the Arabs to China during the Tang Dynasty (618–907). These *Kunlunnus* were fictionalized in different short Chinese stories as talented and skillful, loyal to their masters, and having mysterious magical power.[34] However, this image was marred by the prevalence of the slave trade in the Song Dynasty (960–1279). Although to the Chinese, Africans were still categorized as strong and stalwart retainers, they failed to become as heroic and magical as before. Instead, they gained reputations for being ill-adapted to their Chinese surroundings: they ate raw food, had unintelligible speech and desires, and could only understand, but not speak, Chinese, even after being "domesticated" for a long time. After that, the *Kunlunnus* were no longer considered as humans, but were instead referred to as *shengfan* (savages) or *guinu* (devil slaves).[35] That the Africans were also called *yeren* (wild men) or *heigui* (Negroes) suggests that the perception of "Blacks are equal to slaves" was well-established in the old Chinese mentality.[36] The Chinese once simply characterized Africa as a land of *hundun* (chaos), i.e., at a very primitive stage of civilization.[37]

In the second stage of Sino-African relations, both parties came under the influence of Western colonialism. Chinese people began migrating to Africa to seek a better living, and in the mid-eighteenth century a small Chinatown emerged in Mauritius. At that time, the Europeans found Chinese laborers to be more capable than the Africans, while the Africans were known as being relatively purer morally and less wicked.[38] Even so, a sense of brotherhood gradually developed between Chinese and Africans as they sought mutual support from each other, especially in the international arena to fight against colonialism.[39]

This sense of brotherhood apparently came to the fore after the rise of Communist China in the third stage of Sino-African relations. Suggesting them as being missionary-oriented, Snow argues that the Chinese after 1949 perceived themselves as obliged to give lessons to the Africans, believing that the African colonies or their newly independent nations were simply reenacting the brutal feudal past of China.[40] For instance, as Dikötter recalls, the Chinese once posited themselves in a play performed in Rwanda as the savior of an African throne from European invasion.[41] Coupled with the internationalist doctrine proposed by Chairman Mao, the Chinese once again saw themselves as superior to their "African buddies." The old images of the Africans were thus reinstated.

"Poor," "Lazy," "Sexist," "Threatening": Stereotyping Africans as Inferior Chinese Partners

These Chinese images of Africa portrayed by Snow and Dikötter can repeatedly be found in our research. To start with, there may be on the Internet one or two random postings portraying the romanticized image of Africa: the "sincere faces and soft music," the romance of Casablanca, the spectacular Victoria Falls and the like.[42] However, the almost-overriding perception of Africans that we gathered online from China is that they are generally economically and spiritually poor. Even the sympathetic users stereotype Africans as their "poor buddies" (*qiongxiongdi*), a term coined in the Maoist era. While poverty in Africa is not without factual support, the general online Chinese view that the Africans are "lazy" is more problematically grounded.[43] For example, when commenting on the purpose of the African summits in which China has participated, most Chinese Internet users reiterate that Africans are "lazy and inefficient at their workplaces," and draw a sharp contrast with the industry of Chinese workers.[44] This is also the general impression given by Chinese who claim to have actually visited Africa.[45] Understandably, many Chinese Internet users see illegal African immigrants present in China as "paralyzing" Chinese society by unfairly tapping into the resources of the socialist welfare system.[46] Even the few comments in defense of the Africans are often Social Darwinist-based; they claim that the laziness of Africans can be explained away as "owing to the hot weather in Africa."[47] As follow-on to this logic, those Africans who choose to remain in China illegally are assumed to be doing so because they wish to stay in such a "paradise" forever.[48] The dissentients of poorly educated Africans can always be recognized by the terms they attribute to them, which is the main focus of Diköter's study already discussed.

These described perceptions of Africa are also found in some circulating Chinese publications. However, the extremely politically incorrect sexual references to Africans are more predominantly an online Chinese product, even though the origins of these seem to lie in the old *Kunnunlu* metaphor discussed by Snow. Some Chinese Internet users believe that the physical strength of Africans is an indicator of their having sexual organs of "abnormal size."[49] This is despite the fact that Africans' mystical strength, as viewed in the Tang Dynasty, is not necessarily commented on negatively online. During the Beijing Olympics in August 2008, for example, many online messages expressed amazement at the African athletes, saying things with racial overtones like "it seems that the darkest among the Negroes are the most powerful ones."[50] But the level of sexual misperception becomes much more racist once the topic is raised online. In a message warning fellow Chinese women to "beware of the Africans," various Internet users said:

> The Negroes love to go clubbing. But their desire is not limited to dancing or drinking. Those Negroes will first try to flirt with our Chinese girls, by all possible means to flirt with us. If there had been only Negroes inside the clubs or discos, raping would have happened already.[51] To my fellow female citizens: for

> the sake of your personal safety and health, for the sake of not giving birth to a black baby who will be discriminated against, for the sake of your happy marriage, please don't hang out with a Negro![52]

In a further posting commenting on illegal African immigrants in China, another user warns fellow Chinese girls that the main purpose of the Africans in China is to look for sexual partners:

> Perhaps because their society is relatively open, they love to have sex; and they cannot stand having no sex for more than a week. They will try to flirt with those young ladies, but seldom succeed. Even if they have a stable partner, they will still look for hookers. If they are out of money and girls, they will go for raping![53]

Thus, the allegedly unreasonably active sexual behavior of the Africans, as some Chinese Internet users perceive it, presents a threat to morality and health within Chinese society. For instance, there is a story circulating online that "10 out of 40 Africans who go for medical checks in China are found to have sexually transmitted diseases"; reports that Chinese university students have been infected with HIV by fellow African schoolmates are frequently found.[54] Such alarming statements prompted even the sympathetic Chinese, like the following Guangzhou user, to be alert:

> We should help and develop our friendship with the African buddies. However, this does not mean that we should allow the Negro population to grow so rapidly and unreasonably in Guangzhou. AIDS won't be spread unless there is emancipation.[55]

Worse still, this stereotype is extended to those Chinese who have established close relationships with Africans. Chinese girls who have fallen in love with Africans are often described as "horny," "female liberals" (*nu xiaoyou*), or "too open." Sentiments expressed in the following posting are typical:

> I ran into a girl who was wearing the black uniform of the Inland Revenue Bureau. I found her respectful when I first saw her. But I was shocked when I noticed the person beside her—a Negro! Yes, a Negro who was dressed like a gangster. Oh my God! The lady in uniform, do you want our Chinese to lose face just because you were hanging out with this freak?[56]

Given the above stereotyping of Africans, which clearly reflects attitudes similar to those studied by Snow in the first stage of Sino-African relations, what contribution do Chinese Internet users see them making to China? When commenting on the kidnapping of Chinese in the Niger Delta in 2007, a user offered an innovative option for the Africans; this was to fight for China on inhospitable terrain, like the Indonesian jungle.[57] Another user suggested the Africans would only be useful to China if they could help kill "U.S. spies" in Nigeria, so as to defend Chinese interests in the area.[58] Compared with the superior civic virtue and cultural attributes of the Chinese, the strength and force of the Africans seem to be their only valuable assets in the eyes of online Chinese.[59] In other words, even though the Chinese users do not see Africans in a very positive manner, they still rather wishfully keep certain stereotypical positions for Africans to fill in the international arena—presumably for China's benefit. Interestingly, it seems to Chinese Internet users that Africans have little choice but to cooperate with the Chinese as an inferior, junior partner in the world. This perceived relationship between Chinese and Africans is nothing odd to the Chinese Internet users. It can, to a certain extent in their view, be readily compared with the relationship between the dominating ethnic Han Chinese people (who are perceived as more "civilized") and the ethnic minorities (who are perceived as "barbaric") that existed in the dynastic feudal China of the past. As a result of contemporary pride experienced in China, brought on by recent economic success, this centric-peripheral worldview, with its roots in China's ancient "Middle Kingdom" complex, can still be applied to the present day.[60]

Doomed Disciples with the Same History: The Fatalist Africans Should Learn From the Chinese Experience

Why are these African stereotypes being reinforced within the online Chinese community? One rationale of the users seems to be to employ the African image to highlight China's recent contrasting success. Indeed, most online messages describe China as an ancient nation sharing a similar history with Africa. To them, contemporary Africa is just like the China of old. Equating Africa today with China's dynastic past or the chaos of its Cultural Revolution, the modern-day Chinese can easily enjoy a sense of superiority. This sense of superiority is comparable to the core missionary feature described by Snow in his third stage of Sino-African relations. For instance, when a user comments on the "clash of civilizations" in Africa, the user says:

> Some Africans do not allow photo-taking because of the backwardness of their villages. This is exactly how we behaved during the Cultural Revolution. . . . Behind their pride and arrogance is the deep sense of inferiority in their soul—which is similar to our own contemporary history.[61]

As another example, a sense of timelessness is often projected for Africa:

> I saw many villagers. Most of them were sitting under a tree and chatting with each other. It made me feel like coming back to the 1970s in the Chowyue Village (where I had my earliest memory) . . . but this is Nigeria in 2006.[62]

Sometimes, such historic analogies restrain users from making critical comments about Africans.[63] But they are not in the mainstream. Even though Africa is seen as sharing a similarly traumatic history with China, most Chinese Internet users do not believe that Africa can eventually catch up and emulate the current Chinese miracle. Such a belief is firmly entrenched in the online Chinese stereotype of the African, as discussed in the last section. The "internal problems of the Africans and the African culture" are frequently cited by the Chinese users to account for the "impossibility" of Africa catching up with China.[64] The following message from someone who allegedly had working experience in Africa is frank about the sense of doubt in the ability of the Africans:

> At the time of my arrival in Africa, I truly respected the Africans and showed sympathy toward them, thinking that they were really as friendly as described by the official Chinese discourse. But after I had arrived, I realised I was wrong. The Africans are selfish, witty, dishonest, free-riding . . . they always gave me the impression of being lazy and inefficient. Their society was corrupt and hopeless. I confess that before the opening up of China, we may also have given similarly hopeless impressions to Westerners. But now even the foreigners know that one day Chinese economic performance will surpass the Americans.[65]

Another Internet user who allegedly observed the Chinese training of African military in Nanjing believes the same. To this user, the Africans are all but inferior:

> Every procedure is clear and fast in China. The African fellows feel scared and nervous, thus they wholeheartedly give applause to the Chinese trainers. During training, the Chinese military demonstrated to the Africans their advanced mine-sweeping machines and vehicles, which had been developed in China, shocking fellow Africans over the advancement of Chinese technology.[66]

The sense of disapproval also has roots in what some Chinese users generically term "the African socio-political system." Although the Chinese sociopolitical system does not have full democracy either, to many Internet users, China's authoritarian model seems to be able to offer a perfect substitute for the chaos in Africa:

> To conclude, development in Africa is trapped in a vicious cycle: development—democracy (or dictatorship)—wars and chaos—development—wars and chaos—democracy (or dictatorship). It is just like the history of dynastic change in China when peace and chaos alternated. So how can Africa really advance in economic and social conditions?[67]

Cutting Costs—Low Priority Should Be Given When Money Is Involved: Online Calculation About Realist-Driven Sino-African Relations

Given the above impression of the inferiority of Africa and the stereotypically assumed impossibility that Africans could catch up with the Chinese, either economically or politically, many Chinese Internet users express cynicism and skepticism about developing a closer relationship with African nations, particularly when money is involved. To the users of SNF in particular, the implied benefits the Africans brought in the past, especially during the peak of Sino-African friendship during the Maoist era, have already been dismissed.[68] They see the "real intention" behind African nations' befriending of China as being almost totally interest-driven, rather than looking upon the Chinese as their "real friends."[69]

Since the intentions and the sincerity of the Africans are predominantly interpreted in a realist manner by the online Chinese community, whether Chinese investment in Africa can bring reciprocal realist returns for China is of prime concern for the Internet users. Most of them do not think China's reentry into Africa is of benefit to their nation, at least not in monetary terms. Some argue that Africa is the only continent available for the Chinese to explore "simply because the Westerners have no interest in it at all."[70] Quoting the famous realist doctrine of "no eternal friends but only eternal interests," some users propose "constructing our eternal interests with the US instead of making friends with several poor African buddies," since "both of them are bad guys anyway."[71]

The aid that China gave Africa in the Maoist era remains for many a sore issue. Consequently, discussing China's current aid program, many have expressed doubts as to whether the "huge Chinese aid" given to Africa would be redeemed this time also; they think that China's African campaign is far from being cost-effective.[72] After the Sichuan earthquakes in May 2008, some Internet users bluntly asked: "Which is more important, the huge developmental aid for Africa or the provision of free elementary education for our kids?"[73] In such a context, because of the lack of comprehensive socioeconomic ties between China and Africa at the popular level in society, assisting Africans out of diplomatic and selfish needs, or assisting Sichuan's inhabitants for humanitarian and livelihood needs, is framed under a dialectical dichotomy; selecting a dichotomized answer from the two options poses no difficulties for almost all online Chinese users. In comparison, when natural disasters, such as earthquakes, occur in Taiwan, the same dichotomy is less easily imposed because the humanitarian needs of the Taiwanese are directly tied to their family or business affiliates in mainland China.

Although some of these comments acknowledge the abundant supply of natural resources in Africa, even the most positive comments on China's economic expenditure in Africa only see Africa as a "gold-mining route" for ad hoc individual Chinese pioneers or desperados.[74] The potential realist benefits for China in gaining access to a cheaper energy supply from the continent as a whole, and the strategic importance of Africa to Chinese diplomacy, are not fully attended to by the Chinese users.[75] Most online messages simply equate the current African campaign of China with that of the Maoist era, the uneconomical returns from which still haunt collective Chinese memory. The gains China stands to make in the campaign, as studied by Peter van Ness, are unknown to the users.[76]

Africa "Cannot Be Abandoned": Constructing China's National Pride and Benevolent Identity via Africa

Paradoxically, despite these sentiments, the Chinese do not wish their country to abandon Africa either. If somehow the Chinese are seen as "losing face" in Africa, the Internet users will respond rigorously and complain that Beijing has not done enough. Any negative behavior, or simply negative comments, coming from Africa about China are taken seriously by the Chinese Internet users:

> Originally I believed the Africans had a good impression of the Chinese. Owing to the aid we've offered and our economic development, the Chinese should have their respect. However, my African friend showed nothing toward us but disdain. "The Chinese people disappointed me," she said seriously. I was silent and my heart was broken.[77]

Despite this confession, quite often Chinese users simply deny that Africans might have any bad impressions about the Chinese. For example, when discussing why there had been kidnappings of Chinese in Ethiopia and Nigeria in recent years, some postings expressed the belief that the sole reason for this was that the Chinese workers had been mistaken for Japanese.[78] A more popular belief is that the Western conspiracy of portraying China as an aggressive country in Africa led directly to the kidnappings.[79] The reasons elaborated upon in Western media to account for these abductions, such as being a backlash to the flood of cheap Chinese products into Africa, are largely ignored by the online Chinese community.

When unfriendly African gestures are leveled at the Chinese, the Internet community will mount a certain degree of pressure on the government. For instance, when a Chinese freighter—trying to deliver supplies to the Mugabe regime in Zimbabwe—was prevented from coming into port in South Africa and subsequently expelled by the South Africans (under American pressure) in April 2008, many users were furious. They assumed South Africa to be like any other African nation and used to being befriended by Beijing. Few of them seemed to know that South Africa was highly critical of Beijing and almost established formal diplomatic relations with Taipei in the 1990s. They exclaimed that "even the Africans do not allow our ships to anchor, and bully us," calling this "shameful," and finding it surprising that "even our African friends are not welcoming us now."[80] To most of them, it simply meant that "our African policy is not working well," and they asked government officials to "make a deep reflection" and to go to Africa to sort it out. They also expected "official protests" but none was forthcoming.[81] This example shows that emotionally the Chinese cannot afford to be brushed off by the Africans, whom they perceive as being inferior to themselves in many ways. Without this junior partner and admirer of China reflecting China's relative success, the Chinese users find it hard to flaunt their sense of superiority.

This is why even though many Chinese Internet users have asked the government to reduce the scale of their engagement with the Africans, the Chinese commitment in Africa still gives them a sense of national pride. Although they do not like the Chinese government investing so heavily in Africa, they still hope Beijing will continue some "benevolent policies" in Africa in order to help Chinese nationalists sustain their self-esteem.[82] Indeed, the Chinese users often compare the Chinese aid offered to Africa with that offered by the United States, the European Union, or Japan, claiming the intention of these countries is "never sincere."[83] They believe that the unconditional aid offered by the Chinese "is not another form of colonialism" but grants the Africans "a sense of confidence that cannot be gained from the Europeans."[84] To them, the contribution of China in the development of Africa is more beneficial to the Africans than that offered by the Europeans.[85] Even the United States should "take the Chinese train" if they want to enter the African market.[86] This image of China, which is built upon but goes beyond Snow's categorization, fits quite well with the recent official rhetoric to construct China as a "responsible state" in the global arena.[87] In short, most Chinese Internet users agree with the strategy of constructing national pride by using relations with Africa to project a benevolent image to the world. The issue is simply how to do it in a more economical manner.

Implications of the Virtual Chinese Perception on the African Reality

Granted, the online expression in China about Africa does not represent the perception of the whole nation. One cannot expect Internet users to offer theoretical analyses as scholars do, or to spell out politically correct statements as the government does. At times, what is expressed online is totally different from what the same person may express in real life. However, the Chinese image of Africa expressed online undoubtedly represents a significant portion of overall Chinese popular perceptions about Africans. An analysis of the major characteristics of the perception and how these may in turn influence actual Sino-African relations follow.

To start with, Africans have become convenient straw men through whom Chinese Internet users can project their wished-for Chinese identities. As Gerald Delanty and Chris Rumford have studied, collective identity always concerns "a relation of self and other by which the identity of the self is constituted in symbolic markers."[88] By labelling the Africans as Edward Said's orientalist concept of "the other," Chinese Internet users could reconstruct their own "selves."[89] The identification of the African "other" is indeed a discursive practice, and a mental construction and reconstruction of the definition of the Chinese in the online community. To be more specific, as Tzyetan Todorov suggests, the construction of "the other" can be systematically studied in three dimensions: the axiological level, the epistemological level, and the praxeological level.[90] At the epistemological level, the Africans are constructed as past friends of China, rather than present friends, as the Chinese Internet users always stress how different they themselves are from contemporary Africans. At the axiological level, the Chinese Internet users doubt the purpose of Sino-African relations. At the praxeological level, they express clearly that the emergence of an unfriendly African attitude toward China, or the catching up of the inferior Africans with the superior Chinese, are both unacceptable to such constructed "selves." Once Africa fulfills the imagination of the Chinese Internet users, many side-products will have been created as a result of the construction of the "self" against the "other." Several members of the new generation of Chinese Internet users adopt an online alter ego, to use as a platform for carrying out their public discursive rights. In turn, what they express online may be taken into consideration by the party-state when formulating future Chinese foreign policy.

The lack of attention to China's realist advancement in Africa is the most striking feature of Chinese online expression, as has been noted by Western scholars, for whom China is often depicted as the nation that has gained most in Africa in the recent decade.[91] This phenomenon suggests that the looked-for, unselfish, benevolent image of China in Africa is needed not only to convince the Africans, but also to try to pacify domestic audiences in China in order to grant them psychological comfort. The presence of China in other countries certainly would not receive the same online interpretation. If the above image is deeply entrenched in the Chinese mentality, few internal sanctions would be imposed on any future realist advancement made by Beijing toward Africa. This would lift a great weight from policy-makers in Beijing, should they wish in the future to further develop oil diplomacy with strategic (and often controversial) African nations, like the Sudan.

In addition, our study suggests that the definition of a "responsible state" is very differently interpreted by Chinese Internet users and Western academia. To Western scholars like Rosemary Foot, encouraging China to become a responsible state naturally implies the embrace by China of a set of universal values like human rights, democracy, and concerted security.[92] In contrast, while most Chinese Internet users firmly believe that China has fulfilled its responsibility to Africa in a benevolent and noncolonialist manner, that impression is based primarily on China's materialistic assistance offered to the Africans. Adhering to a set of universal values and pushing them in Africa, to the online Chinese, is simply irresponsible. This suggests that China's policy in African regions like Darfur receives much backing from its online subjects, as these people expect Beijing to play a neutral role without obvious value judgment being applied to the Africans. In short, the "responsibility" the West expects China to fulfill in Africa is ironically seen as "irresponsibility" by the online Chinese community.

Finally, although Chinese aid and investment in Africa are not popularly supported by the Chinese Internet users, a nationalist sense of glory guarantees the continued future involvement of China in Africa. If China becomes more affluent and maintains a strong position during the current global financial crisis, the Chinese users may well be more relaxed about investing in Africa, even though this may imply a financial loss. The existence of the online community may well push Chinese foreign policy toward Africa in an increasingly constructivist-driven manner, with the coining of more high-sounding slogans to satisfy the nationalist expectation of the masses. Instead of satisfying the nationalist ego, as the Maoists did by offering excessive economic assistance without considering national capability, contemporary China will focus more on constructing a righteous global image and a more cautious deployment of investment on the African continent.

However, if China experiences an obvious economic downturn, it is almost certain, as online responses have indicated, that expenditure in Africa would top the list of costs that Chinese would like seen cut. Coupled with the fact that the Beijing government has been cagey about its own initiative in scrambling for resources in Africa, the intrinsic value of China's presence in Africa is likely to be understood by the ordinary Chinese in a deflated manner. If one day, views on Africa become too divergent among the party-state's diplomatic, military, and economic sectors, as well as within public and private companies and the general public at large, the government would find itself subjected to increasing pressure to at least be more transparent about its realist intention in Africa. Failure to do this would result in the online civil society being unwilling to endorse such a campaign.

This said, we have to acknowledge the fact that the current political structure in China makes online discussion one of the only platforms where ordinary citizens can comment on current affairs. As Christopher Hughes puts it, such discourse helps the Chinese Internet users "locate themselves in relation to the official discourse on nationalism by appropriating its themes in order to promote and legitimate a wide range of other discourses with which it can be bound up."[93] In other words, as their largely muted voices on the subject of Sino-African relations in daily life reveal, Chinese Internet users (at least as our case study has shown to date) have expressed little intention of pressuring the party-state to act on their opinions. Even if they had such an intention, without an established civil society and institutionalized democracy, their opinions are unlikely to directly influence the real practice of Chinese foreign policy, other than through the ways mentioned above. The impact of the online Chinese perception of Africa should not, therefore, be exaggerated. Nor, however, can it be casually dismissed or underestimated. It currently performs an irreplaceable role in subtly helping create a national perception of Sino-African relations in contemporary China, which forms part of the overall construction of the state identity of the People's Republic in the twenty-first century. It is essential for Africans to comprehend the overall importance of image building in China's African policy, and that in this Chinese Internet users have played a prominent role in shaping and reshaping their nation's perception of Africa.

Notes

*This chapter is a revised version of the essay originally published in *The Journal of Modern African Studies*, Vol.47, No.3, September 2009, 425–48. The author acknowledges the South China Programme of the Institute of Asia-Pacific Studies, The Chinese University of Hong Kong, for supporting the primary research of this article. The author also acknowledges Mr. Jacky Fung and Mr. Wilson Chan for their assistance in the primary research. All of the online messages on Chinese various forums have been accessed between 10 December and 20 December 2008.

1. China Internet Network Information Center (CNNIC), *Statistical Survey Report on the Internet Development in China*, 2008 [www.cnnic.cn/uploadfiles/pdf/2008/2/29/104126.pdf].

2. The author had thought of extending the timeline to November 2008 in order to have a full two-year time frame in the hope of gathering more post-Olympic

comments. However, from September 2008 onward, discussion on Africa has been very extensively influenced by the Obama factor. This is a totally different topic worth a paper in itself.

3. *Qiangguo Xiequ*, "Information Tree—2007," 2007 [bbs2.people.com.cn/fuwu/dt/hm07.html, on 10.12.2008]. And Tianya Community (TYC), "Mailbox for Advertisement Enquiry," 2008 [article.tianya.cn/2005/ad/forty.html, on 10 December 2008]. And TYC, *Information for Tianya Community*, 2008 [help.tianya.cn/about/about.html, on 10 December 2008].

4. See, for example, Message ID 156285, "Shijie Gailan: Renzai Feizhou (Living in Africa: Overview of the World)," TYC, 27 July 2007, accessed 10 December 2008.

5. Geoffrey Taubman, "A not-so World Wide Web: The Internet, China and the challenges to non-democratic rule," *Political Communication*, Vol.15 (1998), 255–72.

6. Michael S. Chase and James C. Mulvenon, *You've Got Dissent! Chinese Dissident use of Internet and Beijing's Counter-strategies* (Santa Monica: RAND, 2002).

7. Yang Guobin, "The Internet and civil society in China: a preliminary assessment," *Journal of Contemporary China*, Vol. 12 (2003), 461.

8. Jack Linchuan Qiu, "The Internet in China: Data and issues," working paper for the Annenberg Research Seminar on International Communication, 2003 [arnic.info/Papers/JQ_China_and_Internet.pdf. html, accessed on 18.6.2008]. And Yuezhi Zhao, "Falun Gong, identity, and the struggle over meaning inside and Outside China," in Nick Couldry and James Curran, eds. *Contesting Media Power* (Lanham, MD: Rowman & Littlefield, 2003), 209–27.

9. Peter H. Gries, "Chinese nationalism: Challenging the States?" *Current History*, Vol. 104: 638(2005), 251–56; Shih-Diing Liu, "China's popular nationalism on the Internet: Report on the 2005 Anti-Japan network struggles," *Inter-Asia Cultural Studies*, Vol. 7: 1 (2006), 144–55.

10. Brenda Chan, "Imaging the Homeland: The Internet and diasporic discourse of nationalism," *Journal of Communication Inquiry*, Vol. 29: 4 (2005), 336–68.

11. Me Zhou, "The 2005 anti-Japanese demonstration in China: the rise of popular nationalism," in *Chinese Papers Collection Online of University Service Centre for China Studies* (Hong Kong: University Service Centre for China Studies, 2007) [www.usc.cuhk.edu.hk/wk_wzdetails.asp?id=6212].

12. Jack Linchuan Qiu, "The Internet in China: Data and issues," working paper for the Annenberg Research Seminar on International Communication, 2003 [arnic.info/Papers/JQ_China_and_Internet.pdf. html, on 18.6.2008], 16.

13. Mark Warschauer, "Language, identity, and the Internet," in Beth Kolko, L. Nakamura, and Gilbert Rodman, eds. *Race in Cyberspace* (New York: Routledge, 2000), 151–70.

14. Gao Jinyuan, "China and Africa: The development of relations over many centuries," *African Affairs*, Vol. 83 (1984), 248–49; Stephen F. Jackson, "China's third-world foreign policy: The case of Angola and Mozambique," in *The China Quarterly*, Vol.142 (1995), 388–422.

15. C. Alden, "China in Africa," *Survival*, Vol. 47:3 (2005), 147.

16. Alden, "China in Africa," 147–48.

17. Eric Downs, "The Chinese energy security debate," *The China Quarterly*, Vol.177 (2004), 21–41; Eric Downs, "The fact and the friction of Sino-African energy relations," *China Security* Vol.3:3 (2007), 42–68; Ian Taylor, *China and Africa: Engagement and compromise* (New York: Routledge, 2006); and Ian Taylor, "China's oil diplomacy in Africa," *International Affairs*, Vol. 82:5 (2006), 937–59.

18. International Energy Agency (IEA), "WEO 2007 Fact Sheet—China." *World Energy Outlook 2007*, 2007 [www.iea.org/textbase/publications/free_new_Desc.asp?PUBS_ID=1987.html, on 10 December 2008].

19. PRC Economic and Commercial Counsellor's Office (ECCO), "Implications from the Lebanese merchants in Sierra Leone to China," 14 April 2003; PRC Ministry of Foreign Trade and Economic Cooperation (MOFTEC), "The situation, problems and solutions about Nigeria's fishery products," 29 March 2005.

20. H. T. Shi, "Zhongguo Nengyuan Anquan De Qianzai Weixie: Guodu Yilai Maliujia Haixia (Latent threat against China's energy security: dependent on the Malacca Strait)," 2004 [news.xinhuanet.com/world/2004-06/15/content_1526222. html on 17 July 2008].

21. Marc Lanteigne, "China's energy security and Eurasian diplomacy: The case of Turkmenistan," *Politics*, Vol. 27:3 (2007), 147–55.

22. David H. Shinn, "Africa, China, the United States, and oil," *African Policy Forum*, 2007 [forums.csis.org/africa/?p=34.html on 7 December 2008].

23. See, for example, the new press for Premier Wen's trip to Africa in the *China Daily*, 16 June 2006, and President Wu's trip in the same newspaper, 12 February 2007.

24. A. N. Goldstein, H. Reisen Pinaud, and Xiaobao Chen, *The Rise of China and India: What's in for Africa?* (Paris: OCED Development Centre Studies, 2006); Domingos J. Muekalia, "Africa and China's strategic partnership," *African Security Review*, Vol. 13:1 (2004), 5–11.

25. Muekalia, 2004, 5–11.

26. Jianbo Luo, "Advancing China's multilateral diplomacy in Africa," *Contemporary International Relations*, Vol. 16:12 (2006), 11–25; Yongping Zhang, "On new international environment and the formulation of China's African strategy," *West Africa and Asia*, January 2005, 19–25. See also Weizhong Xu, "Beijing summit promotes Sino-African relations," *Contemporary International Relations*, Vol. 17:1 (2007), 72–79.

27. Berry Sautman and Hairong Yan, "Friends and interests: China's distinctive links with Africa," *African Studies Review*, Vol. 50:3 (2007), 75–113.

28. Jack Linchuan Qiu, "The Internet in China: Data and issues," working paper for the Annenberg Research Seminar on International Communication, 2003 [arnic.info/Papers/JQ_China_and_Internet.pdf. html, on 18 June 2008].

29. A. Lorenz, "A technocrat riding a wild tiger," *Der Spiegel*, 10 November 2005.

30. Alexander Wendt, "Anarchy is what states make of it: the social construction of power politics," *International Organization*, Vol.46:2 (1992), 391–425.

31. Douglas Porpora, "Cultural rules and material relations," *Sociological Theory*, Vol.11:2 (1993), 212–19.

32. Philip Snow, *The Star Raft: China's encounter with Africa* (London: Weidenfeld & Nicolson, 1988).

33. Snow, 18–19, 1988; Frank Dikötter, *The Discourse of Race in Modern China* (Hong Kong: Hong Kong University Press, 1992), 16.

34. Philip Snow, *The Star Raft: China's encounter with Africa* (London: Weidenfeld & Nicolson, 1988), 17.

35. Snow, 19, 1988.

36. Frank Dikötter, *The Discourse of Race in Modern China* (Hong Kong: Hong Kong University Press, 1992), 15–17.

37. Dikötter, 50, 1992.

38. Philip Snow, *The Star Raft: China's encounter with Africa* (London: Weidenfeld & Nicolson, 1988), 46.

39. Snow, 66, 1988.

40. Philip Snow, *The Star Raft: China's encounter with Africa* (London: Weidenfeld & Nicolson, 1988), 71.

41. Frank Dikötter, *The Discourse of Race in Modern China* (Hong Kong: Hong Kong University Press, 1992), 194.

42. Message ID 7835502, "Qingxi Feizhou (Emotion Attached to Africa)," DZH, 18 October 2007, and Message ID 142695, "Youyou Yanzhong Di Feizhou Dalu (Traveling to the African Continent)," TYC, 5 January 2008.

43. Message ID 166266, "Wenming Chongtu Zai Feizhou (Clash of Civilizations in Africa)," TYC, 30 December 2007; Message ID 2627, "Ouzohuren (Bairen), Zhongguoren, Feizhouren (European Whites, Chinese, Africans)," AB, 26 May 2007; and Message ID 85548424, SNF, 28 April 2008.

44. Message ID 534, "Feizhou Fenghui (The African Summit)," *Blog Buxu Lianxiang (No Associative Thinking)*, 2 November 2006 [www.wangxiaofeng.net/?p=534.html].

45. Message ID 708, "Xifeiyouji Feizhouren Yinxiang (Journey to West Africa: Impression of the Africans)," GGW, 18 September 2007.

46. Message ID 58305, "Zhongguo Nurenmen, Qing Xiaoxin Heiren Ba (Chinese Women, Please Beware of the Negroes)," *163*, 28 October 2006.

47. Message ID 200781164714, "Shuoshuo Guoren Dui Feizhou Di Wujie (About the Chinese Misconception of Africa)," *Blog of Bo Ku*, 1 August 2007 [www.126blog.com/u5/guleite/archives/2007/200781164714.shtml#].

48. Message ID 84, "Feifa Zhiliu Zhongguo De Feizhouren (Illegal African Immigrants in China)," *KaKu De Buluoge* (Kaku's Blog), 23 May 2007 [www.kakug.com/blog/archives/84. html].

49. Message ID 82043, "Feizhouren Weishenme JJ chang" (Why Do Africans Have Long Dicks), *Hunan Daxue Xiaoyuan Luntan*, 19 March 2008 [wlzb.net/phpwind/read.php?tid=82043. html].

50. Message ID 148612, "Tiyu Jujiao: Tianjing—Ershijiu Jie Aoyunhui Tianjing Bisai Zhibo Taoluntie (Sports Focus: Field—The Discussion Post of Live 29th Olympic Games)," TYC, 18 August 2008, accessed 20 August 2008.

51. Message ID 58305, "Zhongguo Nurenmen, Qing Xiaoxin Heiren Ba (Chinese Women, Please Beware of the Negroes)," *163*, 28 October 2006.

52. Message ID 84, "Feifa Zhiliu Zhongguo De Feizhouren (Illegal African Immigrants in China)," *KaKu De Buluoge* (Kaku's Blog), 23 May 2007 [www.kakug.com/blog/archives/84. html].
53. Message ID 84, 2007.
54. Message ID 58305, "Zhongguo Nurenmen, Qing Xiaoxin Heiren Ba (Chinese Women, Please Beware of the Negroes)," *163*, 28 October 2006; Message ID 86455443, SNF, 28 May 2008.
55. Message ID 86616883, SNF, 8 June 2008.
56. Message ID 58305, "Zhongguo Nurenmen, Qing Xiaoxin Heiren Ba (Chinese Women, Please Beware of the Negroes)," *163*, 28 October 2006.
57. Message ID 2_2104481, "Nirier Panjun Bangjia Zhongguoren Di Liyou Xiangdang Guiyi (Very Weird Reasons for the Niger Insurgents to Kidnap Chinese)," CIB, 13 July 2007.
58. Message ID 85595517, SNF, 3 May 2008.
59. Message ID 2627, "Ouzohuren (Bairen), Zhongguoren, Feizhouren (European Whites, Chinese, Africans)," AB, 26 May 2007.
60. Thomas J. Christensen, "Pride, pressure and politics: The roots of China's worldview," in Yong Deng and Fei-Ling Wang, eds, *In the Eyes of the Dragon: China views the world* (Lanham, MD: Rowman & Littlefield, 1999); and Andrew J. Nathan and Robert S. Ross, *The Great Wall and the Empty Fortress: China's search for security* (New York: W. W. Norton, 1997).
61. Message ID 167971, Response to "Wenming Chongtu Zai Feizhou (Clash of Civilizations in Africa)," TYC, 30 December 2007.
62. Message ID 166266, "Wenming Chongtu Zai Feizhou (Clash of Civilizations in Africa)," TYC, 30 December 2007.
63. Message ID 166266, 2007.
64. Message ID 2627, "Ouzohuren (Bairen), Zhongguoren, Feizhouren (European Whites, Chinese, Africans)," AB, 26 May 2007.
65. Message ID 2627, 2007.
66. Message ID 7918396, "Feizhou Junren Nanjing Xue Saolei (African Military Learns Mine Sweeping in Nanjing)," DZH, 25 November 2007.
67. Message ID 2627, "Ouzohuren (Bairen), Zhongguoren, Feizhouren (European Whites, Chinese, Africans)," AB, 26 May 2007.
68. Message ID 85014172, SNF, 12 March 2008.
69. Message ID 85217128, SNF, 30 March 2008.
70. Message ID 85611130, SNF, 4 May 2008.
71. Message ID 534, "Feizhou Fenghui (The African Summit)," *Blog Buxu Lianxiang (No Associative Thinking)*, 2 November 2006 [www.wangxiaofeng.net/?p=534.html].
72. Message ID 2308022, SNF, 20 May 2007; Message ID 85498640, SNF, 23 April 2008.
73. Message ID 2308022, SNF, 20 May 2007. See also Message ID 86205651, SNF, 13 May 2008; Message ID 86244154, SNF, 16 May 2008.
74. Message ID 809662, "Dao Feizhou Taojinqu (Go to Gold-mining in Africa)," TYC, 20 October 2006; Message ID 85548424, SNF, 28 April 2008.

75. Message ID 85428195, SNF, 18 April 2008.

76. Peter van Ness, "China as a Third-World State: Foreign policy and official national identity," in Lowell Dittmer and Samuel Kim, eds. *China's Quest for National Identity* (Ithaca, NY: Cornell University Press, 1993), 194–214.

77. Message ID 85594156, SNF, 3 May 2008.

78. Message ID 85420541, SNF, 17 April 2008.

79. Message ID 2_2104481, "Nirier Panjun Bangjia Zhongguoren Di Liyou Xiangdang Guiyi (Very Weird Reasons for the Niger Insurgents to Kidnap Chinese)," CIB, 13 July 2007.

80. Message ID 85496103, SNF, 23 April 2008; Message ID 85496042, SNF, 23 April 2008.

81. Message ID 85496103; Message ID 85496042, 2008; Message ID 85498640, SNF, 23 April 2008.

82. Message ID 156285, "Shijie Gailan Renzai Feizhou (People in Africa: Overview of the World)," TYC, 27 July 2007; Message ID 85548424, SNF, 28 April 2008; and Message ID 86932074, SNF, 28 June 2008.

83. Message ID 86541820, SNF, 3 June 2008.

84. Message ID 1469849, "Zhongguo Geiliao Feizhou Ouzhouren Bugei De Zixin (China Gave Africa a Sense of Confidence that Cannot be Gained from Europeans)," *Jianzhi Lantian* (Sword to the Sky Blog), 7 December 2007 [blog.china.com/u/070106/24849/200712/1460849.html]; Message ID 86541820, SNF, 3 June 2008.

85. Message ID 86906864, SNF, 26 June 2008.

86. Message ID 86682113, SNF, 13 June 2008.

87. See, for example, the interview of the PRC's Foreign Minister Yang Jiechi, *People's Daily*, 24 December 2007.

88. Gerald Delanty and Chris Rumford, *Rethinking Europe: Social Theory and The Implications of Europeanization* (London: Routledge, 2005), 51.

89. Edward Said, *Orientalism* (New York: Pantheon Books, 1979).

90. Tzyetan Todorov, *The Conquest of America* (New York: Harper Perennial, 1982).

91. For an extensive review of how Western academics view Sino-African relations, see Daniel Large, "Beyond Dragon in the Bush: The Study of China-African Relations," *African Affairs* 107:426 (2008), 45–61; C. Alden, "China in Africa," *Survival*, Vol. 47:3 (2005), 147–64; and I. Taylor, *China and Africa: Engagement and Compromise* (New York: Routledge, 2006).

92. Rosemary Foot, "Chinese power and the idea of a responsible state," *The China Journal*, Vol. 45 (2001), 1–19.

93. Christopher Hughes, "Interpreting nationalist texts: A post-structural approach," *Journal of Contemporary China*, Vol. 14:43 (2005), 247.

12

Discussions on Sino-Latin American Relations in *Qiangguo* Forums (or the Lack Thereof)

Kai-chi Leung

On May 1, 2007, the Caribbean island nation of Saint Lucia reestablished formal diplomatic ties with Taiwan. As a response, China suspended its diplomatic relations with Saint Lucia four days later. Looking back, China established formal diplomatic ties with the island nation in 1997 and helped them build a new stadium. This event marked the second time Saint Lucia changed sides in the "diplomatic warfare" between China and Taiwan.

The loss of Saint Lucia generated minimal interest at *Qiangguo*,[1] a major bulletin board system in Chinese online communities. The news was first posted by netizen *lianganshikong* in the early afternoon of May 1. Only four posts responded to it.[2] A total of 4,516 posts were logged at *Qiangguo* on the same day, with most of them related to foreign affairs, military, socialism, and other themes common to *Qiangguo*. The "loss" of Saint Lucia, while being very much related to these common themes, did not attract a lot of attention at *Qiangguo*.

This conspicuous silence has an important implication for the study of cyberculture and Chinese politics. While Chinese online communities, in particular their hypothesized roles in political transformation in China, have received an increasing amount of attention lately, most studies are concerned with issues that are of the netziens' interest. For instance, researchers have paid a reasonable amount of attention to public sentiments toward Japan and the United States. In relation to the online discussion about China–Latin America[3] relations, this chapter attempts to look at the relationship between cyberculture and Chinese politics from a different perspective—that is, instead of looking at what is being discussed, this chapter looks at what is not being discussed.

Instead of assuming some sort of a deterministic role for the Internet (either as a vehicle of democratization or nationalism), it is probably more beneficial to consider the possibility that social actors may not make indiscriminate use of all of the potentials offered by a new technology. In short, a process of selection and adaptation is required in technology adoption. This perspective suggests that it is necessary to have a more sophisticated approach in the study of online Chinese nationalism, so that the fallacy of technological determinism can be avoided.

Cyberculture and Chinese Politics

The discussion of cyberculture and politics in China is often centered around its implication to democratization. The possible implications of the Internet to democratization are highly debated. While the optimists trumpet the endless possibilities of the information age, the critics question just how "new" the new world actually is.

It is often assumed that through the promotion of comparatively free exchange of ideas and the possibility for people to interact globally online, the Internet is likely to stimulate participation and free netizens from oversight of authorities. Such a potential freedom is also often assumed to be a likely impetus for political liberalization in the future.[4] Globally speaking, we have seen the use of the Internet and other information and communication technologies in political movements such as the Zapatista revolt in Mexico[5] and the overthrown of the Suharto regime in Indonesia,[6] as well as the downfall of former President Estrada in the Philippines.[7] More recently, researchers have focused on the expanded public participation through web campaigning in U.S. presidential elections since 2000.[8]

This optimistic view is often criticized by the skeptics as being too idealistic, as the new political space created by the Internet and other information and communication technologies is not necessarily free or progressive. Conservative and extremists groups often make use of the Internet for propaganda and fundraising, as well as for the coordination of their "anti-social" activities.[9] For instance, right-wing militias in the United States and racial hate groups in Europe, as well as Islamic fundamentalist movements originating in the Middle East, have all made use of the Internet as a part of their political strategy.

In China, a similar debate on cyberpolitics is also evident.[10] On the one hand, the proliferation of Internet access has provided a new opportunity for dissidents to speak up and connect with one another. Chinese netizens have made use of e-mail lists and bulletin broad systems to build vibrant online communities. Even though access is still largely censored, netizens have made use of innovations such as file sharing software and proxy servers to circumvent regulations.[11] For example, in 2007, netizens in Xiamen, southeast China, used mobile phone text messaging to protest against a plan to build a chemical processing plant there.[12]

That said, the possibilities provided by Internet technologies are often far more diverse (and in some cases more problematic) than the wishful thinking of the optimists. The rise of online Chinese nationalism has shown that instead of building some sort of a cyberutopia, new information technologies may serve as a vehicle for the promotion of xenophobic (and in some cases racist) politics. For example, the wave of anti-Japanese protests in 2005 was backed by the extensive use of online mobilization and coordination.

The two opposite positions above occupy much of the debate on cyberpolitics in China. However, as many followers of China studies have come to realize, the reality is considerably more complicated than either extreme. While the authority considers the technology a potential threat, it is also actively harnessing it for specific political and economic objectives, such as the changing of scope and structure of government processes through "informatization."[13]

Before getting too far into the social implications of new information technologies in China, it is beneficial to remind ourselves that discussion on this topic often risks the possibility of slipping into a form of technological determinism: technology can act independently as a key governing force in society. Both the optimists and the skeptics have the assumption that technologies have an autonomous functional logic and therefore interact within society as independent agents. While they differ in the results of technological innovation, they very much consider humans as little more than objects of techniques. In other words, Chinese society is often reduced to a passive subject of technological processes in the eyes of both the optimists and the skeptics.

Two issues of equal importance should be mentioned here. First, new technologies are also products of social processes. Forms of technological innovation are largely influenced by the social contexts they operate in. The development of many new technologies, such as nuclear weaponry, was supported by certain highly specific social forces. A simplistic cause-and-effect formula between technology and society will therefore be unable to capture their reciprocal and intertwining relationship.

Second, people do not blindly take every possibility provided by new technologies. In most cases, there are a variety of workable solutions to any given problem, while each of these solutions can often tackle a variety of problems. For example, different designs for bicycles, such as the high-bicycle and the safety bicycle, were initiated with different objectives in mind. The "safety" design eventually won out. However, it did not stop people from using it for certain unintended purposes (such as the invention of BMX). A process of selection and adaptation, shaped by the local contexts, is always necessary in technology adoption.

The emphasis on social context in the adoption of the Internet is not new,[14] yet the question of selection and adaptation is of particular importance in the study of online Chinese nationalism. The rise of nationalism in China coincided with the proliferation of Internet access in the 1990s. If we adopt the optimistic view suggested above, possibilities of meeting people from outside China and gaining information about the world from a wide variety of sources should in principle serve as an impediment to the rise of nationalism. As people get to explore the online world and experience the diversity of the world community on their own, essentialist doctrines drummed up together with nationalistic sentiment are likely to be challenged.

The existence of strong nationalistic sentiment in Chinese online communities hints that the above view does not cover the whole of the technology-society problem. The question of selection and adaptation places doubt on whether netizens are interested in all of the new prospects provided by the Internet. Even though it is possible for Chinese netizens to access information globally from a variety of sources, they may choose to disregard perspectives they are not familiar with. Even worse, they may judge these new perspectives through the lens of their preexisting positions. In such a scenario, instead of providing new social possibilities, the virtual world is subsumed as an extension of the real world, as preexisting controversies and disputes have found new battlegrounds online.

The following study of netizens' views on Latin America at *Qiangguo* will demonstrate the importance of the question of selection and adaptation in technology adoption. The case of Latin America serves as an important test of the selective usage of the Internet among Chinese netizens. From an optimistic perspective, even though Latin America is the farthest continent from China, Chinese netizens should be able to make use of the Internet to gain a more comprehensive view of it. Unfortunately, such an assumption was found to be overly idealistic. Technology adoption, in this case—the use of the Internet for discussion about Latin America—was found to be a rather biased process. As suggested by the question of selection and adaptation, Chinese netizens did not embrace all of the possibilities equally.

This chapter will tackle the question of selection and adaptation in three steps. First, it will discuss the contemporary relationship between China and Latin America. Second, it will look at the Chinese netizens' limited understanding on Latin America. Third, it will discuss in what ways the netizens have failed to appreciate the multifacetedness of China–Latin America relations. Finally, it will conclude with an inquiry on whether the biased portrayal of Latin America online may provide insights for further investigation in cyberculture and Chinese politics.

China's Interest in Latin America

To demonstrate that the netizens' understandings of Latin America are limited, a quick review of China's interest in Latin America is necessary. The following review is drawn primarily from academic and policy papers on Sino–Latin American relations. It shows China's interest in Latin America is multifaceted, even though it is not always acknowledged by some of the Chinese netizens.

While China's growing interest in Latin America is often considered as a recent phenomenon,[15] they actually have a very long history of exchange. Trade between China and Latin America dates back to the Age of Discovery, often through the Manila Galleon Route. Today, large and vibrant Chinatowns are thriving in Latin America, such as those of Mexico City, Havana, and Lima (figure 12.1). These Chinese communities started with the rise of Chinese emigration in the nineteenth century to various countries in Latin America. Most migrants came from Guangdong Province and worked as contract laborers in primary production such as farming and mining. Many of them also contributed to construction of infrastructure, such as the Panama Canal.

Regarding the official connections, cooperation among China and countries in Latin America was highlighted by Mao Zedong in his aspiration of uniting the people of Asia, Africa, and Latin America (*yafeila*) against Western imperialism. As Mao put it, "This is the fundamental principle of people in Asia, Africa, and Latin America united against imperialism. Our mutual relationship is a brother's relationship, not a father-to-son relationship."[16] More recent visits by senior officials are less concerned with the building of a Third World coalition, but are largely motivated by economic incentives, as demonstrated in Jiang Zemin's visit in 2001, Hu Jintao's visit in 2004, and Zeng Qinghong's visit in 2005.

China's primary economic interest in Latin America is access to resources. Importation of raw materials such as oil, metals, lumber, and agricultural products has been growing quickly.[17] China's imports from Latin America grew more than 600 percent from 1999 to 2004, from almost $3 billion to $21.7 billion.[18] China is also interested in exportation to Latin America. Exports are not limited to apparel traded by small and medium enterprises, but also capital and technology intensive products such as turboprop-powered airliners made by the China Aviation Industry Corporation.

FIGURE 12.1
Chinese Newspapers in Lima's Chinatown (Barrio Chino)

From a political perspective, the increasing influence of China in Latin America may contribute to further diplomatic isolation of Taiwan. As of Feburary 2009, the Republic of China maintains diplomatic relations with twenty-three countries in the world, of which twelve are in Latin America and the Caribbean. China has been actively competing against Taiwan for political influence in the region with its increasing economic strength. The use of multimillions aid packages by both sides for consolidation of diplomatic ties is often referred to as "checkbook diplomacy."[19]

Speaking of international relations, China's growing interest in Latin America is often considered a threat to United States influence in the region.[20] It should be noted that Chinese involvement in Latin America so far is still very marginal when compared with the well-established U.S. economic and political linkages. The former George W. Bush administration tended to downplay the potential threat of China and insisted that China's influence in the region was minimal.

Perception of China's growing interest in Latin America varies in the region. A report from the Inter-America Development Bank listed four dimensions of significance: China as an example of success, as a market, as a partner, and as a competitor.[21] First, the rapid economic growth of China may serve as a potential source of policy lessons. Many countries in the region have referenced their comparatively lagging economic performance with the sustained fast growth of China. Second, the participation of China in the global marketplace provides a fresh opportunity for Latin America's export industries. Rapid growth in China is often considered a major cause of rising primary commodity prices and an important chance for Latin America's export industry to expand. Third, increased economic exchange between China and Latin America may lead to further multilateral cooperation in international forums. The formation of G-20 and the participation of Chinese peacekeeping police in Haiti are two important examples of multilateral partnership. Finally, policy-makers in Latin America have feared their markets may get flooded by Chinese products while foreign investors may leave Latin America and move to China. It should be noted that this notion of China as a competitor is not well supported by the currently available evidence. For example, researchers have found that the growth of foreign investment in China did not significantly affect Latin America's FDI in general.[22]

In summary, China's relationship with Latin America is rich and diverse. China's interest in Latin America is growing, particularly in economic issues such as resources acquisition. China's growing influence in the region has led to questions on its political implication. Political leaders in Latin America are well aware of these questions and have expressed both enthusiasm as well as reservation.

Latin America as a Discussion Topic for *Qiangguo* Netizens

This section attempts to look for empirical observations related to netizens' views on Latin America. Before we get into the observations, a few methodological issues should be highlighted first. The study of cyberculture is so diverse that it is often considered "too undisciplined to be called a discipline."[23] One of the many causes of this heterodoxness is the problem of method. Research on cyberculture is highly susceptible to methodological issues such as sampling and subjectivity.[24] Research on online Chinese nationalism, like this one, is almost inevitably constrained by the lack of an established approach of inquiry. In particular, the question of fractured identities in cyberculture is of paramount importance here. Online Chinese nationalists, for example, is a very ill-defined research subject. The term *fenqing* (angry youth) is more a label for certain online behaviors than for a specific group of netizens. The same person may act very much like a *fenqing* in one discussion but not another.

Instead of attempting the probably impracticable task of locating a representative group of online Chinese nationalists and conducting a focused inquiry on their views, this research takes a more general approach and surveys how topics related to Latin America are referenced in a Chinese bulletin board system. *Qiangguo* is selected here for its reputation as a major bulletin board system in China.[25] It also served as a research subject in a similar study on online Chinese nationalism before and therefore provides the possibility of a longitudinal comparison.[26] Concerning the issue of censorship, with the exception of certain highly sensitive topics such as the June Fourth Crackdown, discussion at *Qiangguo* seems to be generally unmoderated. Posts that directly criticize public policies of the Chinese government are very common.

This research focuses on references to Latin America or related terms in posts logged at *Qiangguo*'s forums. Using the search engine provided in the *Qiangguo* website,[27] all posts from midnight of January 1, 2007, to midnight of January 1, 2008, were examined. Posts from all of the forums listed in the search engine were included. These posts were searched for matches of terms related to Latin America in their titles. These terms include major geographical locations, political leaders, and historical figures in the region. Common Chinese translations in simplified Chinese were used as inputs. When more than one common translation was available, they were searched separately with the results added together. Table 12.1 shows the number of posts found for each of the terms requested. These counts are not exclusive to one another. For example, a post on the revival of socialism with Venezuela, Cuba, Hugo Chavez, and Fidel Castro in its title was counted once in each of the respective searches.

A first-order observation one can make from these results is the lack of posts related to Latin America. The number of posts with titles related to Latin America amount to only a few hundred, while posts with the United States in the title amount to over fifty thousand. Comparatively speaking, there is a general lack of interest in Latin America among *Qiangguo* netizens.

A closer look at the results reveals some interesting tendencies. For geographical terms, countries with larger geographic areas (such as Brazil and Argentina), and countries with a socialist ruling regime (such as Cuba and Venezuela) were more likely to be referenced. For people's names, in general historical socialist figures (Fidel Castro and Che Guevara) received far more posts than newly elected socialist leaders (Lula da Silva and Evo Morales). Hugo Chavez served as an exception here. While his involvement in international politics is comparatively recent, he was mentioned even more frequently than Fidel Castro. While these posts varied in the contexts they were in, many of them were comments on Venezuela "oil socialism" and remarks made by Chavez on U.S. politics.

The following analysis will discuss the thematic interests expressed in the posts in three sections. The first section will be on the two most important themes of discussion among the posts. The second section will look at other

TABLE 12.1
The Frequency of Latin America–Related Terms Mentioned in *Qiangguo* Posts in 2007

Terms	*Search Input*	*Number of Posts*
Latin America	*lamei*[1]	224
South America	*nanmei*	135
Cuba	*guba*	541
Brazil	*baxi*	299
Venezuela	*weineiruila* or *weiguo*	287
Mexico	*moxige*	162
Argentina	*agenting*	126
Chile	*zhili*	63
Peru	*milu*	46
Panama	*banama*	41
Colombia	*gelunbiya*[2]	38
Saint Lucia	*shengluxiya*	19
Uruguay	*wulagui*	13
Suriname	*sulinam*	7
Paraguay	*balagui*	3
Guyana	*guiyana*	3
French Guyana	*fashu guiyana*	2
Caribbean	*jialebi*	54
Falklands	*fukelan or madao*	20
Amazon	*yamasun*[3]	10
Puerto Rico	*boduolige*	2
Zapatista	*sapata*	0
Hugo Chavez	*chaweisi*	250
Fidel Castro	*kasiteluo* or *kasichu*	141
Lula da Silva	*lula*	4
Evo Morales	*molalaisi*	1
Che Guevara	*gewala*	96
Japan	*riben*	28,313
United States	*meiguo or meilijian*	55,675

1. A search on *ladingmei* rendered zero results.
2. Results related to Columbia University (in New York City), Space Shuttle Columbia, and Columbia Pictures were excluded.
3. Results related to Amazon.com were excluded.

significant themes of discussion that were found. The third section will discuss two important issues that are important in real world Sino–Latin American relations, but were largely ignored by the netizens.

Two Major Themes

A closer examination of the content of the retrieved posts revealed several themes of discussion. The two most important themes were economic policies and the revival of socialism, and international relations.

Economic Policies and the Revival of Socialism

By far the most frequented discussion topic on Latin America was its economy. Latin America's economy was often depicted by the netizens as rotten, corrupted, and chaotic. Debates on economic policies of Latin American countries were generally framed as "lessons for China." When netizens tried to make arguments against certain economic policies in China, failures of similar policies in Latin America were often cited to boost their positions.

References to Latin America (*lamei*) were in many cases made in the form of *lameihua* (Latin Americanize). A number of posts warned of the *lameihua* tendency of certain economic policies in China. The precise content of *lameihua* was rarely specified. The term was used in various contexts, including financial crisis, neoliberalization, privatization, inequalities, and polarization. In general, the term was used in a dismissive manner, often as a criticism against current problems in China. For example, with respect to social polarization in China, netizen *wodepipingguan* suggested,

> Is it heading to success? It is being Latin Americanize, poor people are getting poorer, rich people are getting richer.[28]

Latin America was not always discussed as a development paradigm to be avoided. Some posts referred to Latin America's recent revival of socialist-leaning politics very favorably. The success of regional leftist leaders in the past ten years[29] served as an important discussion topic. These posts were largely not concerned with the specificity of the revival of socialist-leaning domestic politics in Latin America countries. Discussion rarely touched on the pros and cons of particular policies such as nationalization of the oil industry. They were more interested in whether the revival of socialism in Latin America should serve as an example for China or not. In an attempt to draw a parallel between socialism in China and Latin America, netizen *yantzejian* borrowed the lyrics of "The East is Red" and stated:

> The East is Red, the sun is rising. South America has brought forth Hugo Chavez. He amasses fortune for the people. Hurrah, he is the people's great savior.[30]

The above positive and negative references worked together and created a perception that people in Latin America regretted their "capitalist past," and decided to embrace socialism as their new development paradigm. Certain netizens cited the revival of socialism in Latin America as an indication that China should avoid the "painful period of neoliberalism" many countries in the region had experienced, and should continue to engage in (a form of) socialism. For example, netizen *youke* suggested,

> The "poor presidents" of many Latin American countries are promoting socialism, building state-owned enterprise, confiscating land, promoting social security. They are learning from Chairman Mao. The developed capitalist United States is right next to them, [how come] they look down to her?[31]

Here, socialist nostalgia in China blended with socialism in Latin America and formed a sense of an upcoming global comeback of socialism. Some netizens even went as far as speculating future timelines of a global socialist revolution, with regions around the world embracing socialism one after another.

An important absence in the discussion of the socialist revival is the lack of posts on the Zapatista struggle in Mexico. While the movement considers itself a part of a wider anti-globalization, anti-neoliberalism social movement, it was not mentioned at all at *Qiangguo*. Also not mentioned is the presidency of Evo Morales in Bolivia, even though he pursuited nationalization of major resources in the country. Given the netizens' interest in socialist revival in Latin America, the absence of posts related to the Zapatista struggle and the Morales presidency is rather striking.

International Relations

Qiangguo netizens expressed a strong interest in foreign policies of countries in Latin America. However, the topics of interest were not diverse. Most of the discussion centered around Latin America's relationship with the "imperial powers," in particular the United States. Other topics in international relations, such as South-South cooperation, were virtually not mentioned at all.

Foreign policies of Latin American countries were understood as either pro-U.S. or anti-U.S. The anti-U.S. leaders were mostly praised as heros. Fidel Castro's disapproval of U.S. politics served as a frequent topic of interest, for instance. Meanwhile, social problems in certain Latin American countries were often linked to their close relationship with the United States, with some sort of a causal relationship implicitly assumed. For example, netizen *tuierjinsan* questioned that

> Iraq and Afghanistan are the most pro-US, with hundreds of thousands of US troops in their countries, did they prosper? Haiti has been pro-US for more than a hundred years, how come it has turned into a UN trustee? Argentina is pro-US, how come it got bankrupted?[32]

The above assertion can hardly withstand any vigorous academic examination. For example, the poor economic condition of Afghanistan has a long and complicated social and historical context, while the direct intervention of the U.S. military in the country is rather recent. That said, assertions like the one above nevertheless reflect a common perception of the United States as a major threat to the prosperity of Latin America.

Similar to socialism, topics related to international relations were often framed as "lessons for China." These "recommendations" on international relations were slightly more subtle than those related to economic policies, yet they nevertheless hit some of the most heated issues in Chinese politics. For example, the Falklands War was cited from time to time to highlight the danger of invoking popular nationalistic sentiment in international relations. As suggested by *luntanbanyungong*,

> Argentina used its rising nationalistic sentiment to start a war against Britain. It ended with the war's initiators being forced to step down in front of the people of the world.[33]

Despite the fact that topics related to imperialism, colonialism, sovereignty, and identity politics were frequently discussed at *Qiangguo*, only two identical returns with Puerto Rico in its title were found, and both contain a picture of Jennifer Hawkins, the winner of Miss Universe 2006. Given the netizens' concern for struggles against the United States in other parts of Latin America, one might expect the struggle of Puerto Ricans for their political future would be of interest to them. However, the political status of Puerto Rico was not a topic of interest at *Qiangguo*.

Another noticeable absence of coverage on foreign relations is the lack of posts on China's involvement in the United Nations peacekeeping mission in Haiti. Only two posts were found on the issue. The first one was a news clipping about the safe return of the fourth peacekeeping team from China. The second one was a response to another post on the Iraq War (on the issue of judicial territoriality). The implication of China's military involvement in other countries to the policy of noninterference, and the fact that Haiti has continued to maintain its diplomatic relations with Taiwan even when Chinese forces were on its ground, were entirely ignored by the netizens.

Other Significant Presences

While the two themes mentioned above covered most of the posts on Latin America–related subjects, there were also a small but significant number of posts on other topics. The following discussion will illustrate three significant presences in discussion topics.

Sports and Local News

Posts related to Latin America were not limited to politics. Netizens also commented on sports and local news. The types of news stories that were of interest to the netizens were rather specific. Virtually all of the posts related to sports were about soccer, while most of the local news posts were reports of violent crime.

As a popular game in both China and Latin America, soccer received a fair amount of discussion. The Chinese male soccer team was mentioned in a number of posts in comparison with various national soccer teams in Latin America. Split opinions were expressed among the netizens on the international standing of the Chinese team. Other popular sports in Latin America (such as baseball in the Dominican Republic, Puerto Rico, and Cuba, and cricket in the West Indies) were essentially not mentioned at all. A pausible explanation of their absence is the general lack of interest in these sports in China.

Coverage of local news in Latin America was limited to mostly news clippings of violent crime, in particular sensational stories related to drugs, sex, and bloodshed. For example, the post "Female inmate was forced to trade sex for food; Prison Scandal Shocked Brazil,"[34] as well as a news article from Xinhuanet, reported that a teenage girl was forced to have sex with multiple fellow inmates in a prison. Another post on Colombia attached a series of pictures of a drug-sniffing dog named Agata, and highlighted that drug traffickers were willing to pay ten thousand U.S. dollars for the dog's head.[35] These posts were mostly adopted from mainland news sources such as Xinhuanet.

Natural Scenery, Pretty Women, and Caribbean Pirates

A small but significant number of picture posts related to Latin America were found. The types of pictures posted were rather specific—scenery (mostly natural landscapes such as beaches) and portraits (mostly young women in swimsuits). A small number of posts on "rare animals from the Amazon" were also found. In the post "The Most Beautiful Brazilian Babe,"[36] two pictures of a female model—one in underwear and another in a bikini swimsuit—were accompanied with an expressive textual descriptions of her "seductive body curves." Particular emphasis was made on her "natural breasts." Meanwhile, many of the posts with the word "Caribbean" in their titles were about the movie series *Pirates of the Caribbean.* The participants were particularly concerned with whether the pirate character played by the Hong Kong movie star Chow Yun Fat was a disgrace for Chinese. Comparatively, fewer posts referred to the Caribbean as a geographic region.

The recurrence of posts on natural scenery, pretty women, and Caribbean pirates, together with the netizens' interests in soccer and violent crime suggested before, point to a question of the roles of imagies promoted by traditional media. Anecdotally speaking, an imaginative geography of Latin America as a remote exotic place is widely circulated in both traditional and online media. Previous research has suggested an interaction of discussion topics in state media news reporting and Internet chat rooms.[37] It is possible that the imaginative geography found at *Qiangguo* is a transplant of a similar one in traditional media. A more specific analysis is needed to detail the formation and transplanting mechanisms of these images.

Meetings Between Chinese Leaders and Latin American Leaders

For a number of small Latin American countries, their names were mentioned at *Qiangguo* only when their leaders met with a Chinese leader. For example, all of the seven posts on Suriname were reports on diplomatic meetings of political leaders from China and Suriname. They were originated from news articles copied from Xinhuanet and Chinese Central Television, with no other comments attached, and no responses followed. These posts were contributed by a small number of netizens who habitually post news articles from Xinhuanet and other Chinese news sources. Posts related to countries and political figures who are lesser known in China would not be mentioned at all without the work of these netizens. That said, the limited variety of news sources cited by these netizens is noteworthy.

Major Absences

With reference to Sino–Latin American relations in the real world, two major absences were found among the retrieved posts.

Trade and Industry

Posts related to international trade between China and Latin America were rare. A handful of posts was found on Chinese exportation of technological intensive goods (such as airliners) to Latin America, and acquisition of natural resources (such as food and fossil fuel) in the region. Discussion about firsthand experience of Chinese doing business in Latin America was not found. Discussion about people from Latin America doing business in China was also not found.

Taiwan

Posts related to both Taiwan and Latin America were rare. Paraguay, the largest country with diplomatic relations with Taiwan, was mentioned in three posts (with two of them related to Taiwan). Other countries with diplomatic relations with Taiwan were sporadically mentioned. For example, when President Chen Shui Bian of Taiwan visited Nicaragua in January 2007, a news clipping about Fidel Castro once urging Nicaragua to end its diplomatic relationship with Taiwan was posted. As mentioned in the beginning, even the declaration of Saint Lucia on May 1, 2007, to establish diplomatic relations with Taiwan did not generate a lot of discussion on the topic. Netizens expressed frustration with the ongoing competition between China and Taiwan with "checkbook diplomacy," but the discussion was hardly a heated one. Considering the lack of posts related to Taiwan, *Qiangguo* netizens did not express a lot of concern with the political influence of Taiwan's presence in the region.

Selective Use of the Internet

The general knowledge on Latin America among the netizens surveyed, as demonstrated by the limited diversity of topics they discussed, was found to be rather biased. They failed to recognize the multifacetedness of Sino–Latin American relations. However, many of them assumed some sort of an expert position on Latin American affairs for themselves, and used issues in the region to support their positions in Chinese politics. It is not evident that the netizens have made extensive use of the Internet as a tool to broaden their limited knowledge on Latin America before they talked about issues related to the region.

China's and *Qiangguo* netizens' interests in Latin America were very different. The recent surge of Chinese investment in Latin America was largely ignored by the netizens. Competition between China and Taiwan for political influence in the region also did not attract much discussion. If one only read the posts on *Qiangguo* for information about Latin America, he or she may not get to know anything about the life of Chinese diaspora and businesspeople in the region at all.

Most of the discussion at *Qiangguo* centered around the recent revival of socialism in the region. Some netizens even called for some sort of a global struggle against imperialism, globalization, and neoliberalism (these concepts were used interchangeably by many *Qiangguo* netizens). However, these calls were so superficial that they focused mostly on well-known figures such as Fidel Castro and Hugo Chavez, and ignored other important actors, including the Mexican Zapatista movement and the Puerto Rican independence movement.

When issues in Latin America were invoked by the netizens, they were mostly used as a reference to contested issues in China. This tendency should be put into the context of the ongoing debates between the "neo-left" and "neo-right" at *Qiangguo*. The two groups accuse one another with derogatory names such as *xiaozuo* (little left) and *xiaoyou* (little right). Latin American countries and political figures were often found to be mentioned only when the netizens wanted to use them to score a point. Successful or failed economic policies in Latin America were discussed mostly to glorify or discredit corresponding contested political stances in China (generally without their relevancy thoughtfully explained). Historical economic disorders in the region, for instance, were attributed to capitalism, imperialism, dictatorship, and many other (often contradictory) explanations, depending on the political stances of the netizens involved in that particular discussion. In short, topics related to Latin America were not cited for discussion directly related to Sino–Latin American relations, but for debates mostly concerned with China's domestic public policies.

What can we make out of these selective and in many instances biased portrayals of Latin America at *Qiangguo*? It should be noted that *Qiangguo* represents a rather specific face of the Chinese online communities. For issues in Sino–Latin American relations that were not well covered at *Qiangguo*, thousands of related Chinese web pages are readily available in other sites. A quick search of "China" (*zhongguo*), "Latin America" (*lamei*), and "investment" (*touzi*) with a search engine like Google rendered millions of results. They include news clippings from both state and foreign media coverage on Chinese business investment in Latin America. Similarly, web pages about Chinese diplomatic relations with Latin American countries, and even first-hand accounts from Chinese diaspora and businesspeople in Latin America, are readily available on the Internet. The problematic selective portrayals of Latin America by *Qiangguo* netizens are not a general phenomenon among Chinese online communities. They represent a peculiar way that the Internet is used among *Qiangguo* netizens.

Here we move back to the question of whether or not the evolution of cyberculture in China involves a process of selection and adaptation in technology adoption. It is clear that at least at *Qiangguo*, a rather selective usage of the Internet was found. The availability of the technology failed to promote more informed and vigorous civil debates.

Reflecting on the Study of Online Chinese Nationalism

Nationalism, likes other forms of discourse, includes a series of representations, practices, and performances wherein meanings are produced, connected, and legitimized. Paying attention to both presences and absences in the texts is consequently crucial for the investigation of the particular forms it takes. It is, therefore, necessary to go beyond the study of the prominent and immediat effects of online Chinese nationalism. It is in this spirit that this research looked at not only how the Internet is being used, but also how it is *not* being used.

Nationalism involves not only the defining of oneself, but also the social construction of "the Other." Henceforth, theorization of online Chinese nationalism must include a detailed analysis of its imaginative geographies of the world. In particular, attention should be paid to the processes of creation and maintainence of these imaginative geographies. This research attempted to shed some light on these processes through calling attention to the clear neglect of the "real" Latin America by certain Chinese netizens.

With this notion of mutual shaping between technology and society in mind, the study of online Chinese nationalism should attempt to look at the variety of ways netizens made use of the technologies. As suggested above, *Qiangguo* netizens represent only a facet of the Chinese online communities. It is possible that both the optimists and the skeptics are correct, that democratization and nationalism are being promoted online at the same time. The observations documented in this chapter are not a fixed set of direct effects of Internet popularization, but results of specific ways in which the Internet is being used and not used by a particulat group of netizens in a diverse world of Chinese online communities.

The above discussion on selection and adaptation highlighted the danger of technological determinism in the study of cyberculture. The relationship between the introduction of a new technology and its practical usage is not straightforward, but reciprocal and convoluted. Chinese society should not be considered a passive recipient of technological innovation. Put in the context of online Chinese nationalism, netizens may not necessarily change their behavior and become more or less nationalistic as they go online. Instead, they select from a variety of possibilities available to them online, adapt these possibilities into a range of practical uses, and create a mosaic of cyberculture in the process.

Finally, with respect to the science and technology literature, the next step of inquiry is arguably an investigation of whether there is anything uniquely "Chinese" about online Chinese nationalism. The problem of biased understanding as described above is not restricted to Chinese netizens. For example, while 22 percent of Americans could name all five Simpson family members in *The Simpsons* animated sitcom, only 0.1 percent of them could name all five First Amendment freedoms,[38] even though the U.S. Constitution is readily available both online and offline. The question of selection and adaptation in technology adoption, and its implication to cyberculture and politics, are probably not limited to China.

Notes

1. bbs.people.com.cn.
2. bbs.people.com.cn/postDetail.do?view=1&id=2159441. Last retrieved on March 5, 2009.
3. For the purpose of this paper, Latin America is defined as a region that consists of Central America, the Caribbean, and South America.
4. Peter Ferdinand, "The Internet, Democracy and Democratization," *Democratization* 7:1 (2000): 1–17.
5. Harry Cleaver Jr., "The Zapatista Effect: The Internet and the Rise of an Alternative Political Fabric," *Journal of International Affairs* 51:2 (1998): 621–40.
6. David Hill and Krishna Sen, "The Internet in Indonesia's New Democracy," *Democratization* 7:1 (2000): 119–36.
7. Cecilia Alessandra Uy-Tioco, "The Cell Phone and Edsa 2: The Role of a Communication Technology in Ousting a President," Fourth Critical Themes in Media Studies Conference (New York: New School University, 2003). mason.gmu.edu/~cuytioco/papers/edsa2cellphones.pdf. Last retrieved on June 3, 2009.
8. Andrew Williams and John Tedesco, ed., *The Internet Election: Perspectives on the Web in Campaign 2004* (Lanham, MD: Rowman & Littlefield, 2006)
9. Phyllis Gerstenfeld, Diana Grant, and Chau-Pu Chiang, "Hate Online: A Content Analysis of Extremist Internet Sites," *Analyses of Social Issues and Public Policy* 3:1 (2003): 29–44.
10. Michael Chase and James Mulvenon, *You've Got Dissent! Chinese Dissident Use of the Internet and Beijing's Counter-Strategies* (Santa Monica, CA: RAND, 2002).
11. Jonathan Zittrain and Benjamin Edelman, "Internet Filtering in China," *Internet Computing* 7 (2003): 70–77.
12. "Middle Class Protest in China," *The Economist*, June 21, 2007, www.economist.com/world/asia/displaystory.cfm?story_id=9367055. Last retrieved on March 5, 2009.
13. Shanthi Kalathil and Taylor Boas, *Open Network, Closed Regimes: The Impact of the Internet on Authoritarian Rule* (Washington, DC: Carnegie Endowment for International Peace, 2003).
14. Helen Milner, "Digital Divide: The Role of Political Institutions in Technology Diffusion," *Comparative Political Studies* 39 (2006):176–99.
15. Kerry Dumbaugh and Mark Sullivan, *China's Growing Interests in Latin America* (Washington, DC: Congressional Research Service, 2005).
16. Speech given at Pyongyang on July 9, 1964, titled "The Prospect of Asian, African and Latin American Struggle from a Historical Perspective," *Selected Works of Mao Tse Tung* (Beijing: Foreign Language Press).
17. David Zweig and Jianhai Bi, "China's Global Hunt for Energy," *Foreign Affairs* 84:5 (2005): 25–38.
18. Kerry Dumbaugh and Mark Sullivan (2005), 2.
19. Daniel Erikson and Janice Chen, "China, Taiwan and the Battle for Latin America," *Fletcher Forum of World Affairs* 31:2 (2007): 69–89.

20. Stephen Johnson, "Balancing China's Growing Influence in Latin America," *Backgrounder 1888*, October 24, 2005 (Washington, DC: Heritage Foundation): 1–6; Peter Hakim, "Is Washington Losing Latin America?" *Foreign Affairs* 85:1 (2006): 39–53.

21. Robert Devlin, Antoni Estevadeoradal, and Andrés Rodríguez-Clare, eds., *The Emergence of China: Opportunities and Challenges for Latin America and the Caribbean* (Washington, DC: Inter-America Development Bank, 2005).

22. Alicia Garcia-Herrero and Daniel Santabarbara, "Does China have an impact on foreign direct investment to Latin America?" *China Economic Review* 18 (2007): 266–86.

23. David Bell, *Cyberculture Theorists: Manuel Castells and Donna Haraway* (London: Routledge, 2007): 52.

24. Norman Denzin, "Cybertalk and the method of instances," in *Doing Internet Research: Critical Issues and Methods for Examining the Net*, ed. Steve Jones (Thousand Oaks, CA: Sage Publications, 1999).

25. This reputation is further reinforced by Hu Jintao's high-profile visit to the site in June 2008. Hu claimed that he visited the site frequently to "listen to the people."

26. Li Xinguang, Xuan Qin, and Randolph Kluver, "Who is setting the Chinese agenda? The impact of online chatrooms on party presses in China" In *Asia.com: Asia Encounters the Internet*, ed. Kong-Ching Ho, Randy Kluver, and Kenneth C. C. Yang (London: RoutledgeCurzon, 2003), 143–58.

27. bbs.people.com.cn/quickSearch.do. Searches last verified on March 4, 2009.

28. bbs.people.com.cn/postDetail.do?view=1&id=2301851. Last retrived on March 5, 2009.

29. Left-leaning Latin American leaders elected in recent years include Hugo Chavez in Venezuela, 1999; Richardo Lagos in Chile, 2000; Nestor Kirchner in Argentina, 2003; Lula da Silva in Brazil, 2003; Tabare Vazquez in Uruguay, 2005; Evo Morales in Bolivia, 2006; and Fernando Lugo in Paraguay, 2008.

30. bbs.people.com.cn/postDetail.do?view=1&id=2071819. Last retrieved on March 5, 2009.

31. bbs.people.com.cn/postDetail.do?view=1&id=1886129. Last retrieved on March 5, 2009.

32. bbs.people.com.cn/postDetail.do?view=1&id=3168849. Last retrieved on March 5, 2009.

33. bbs.people.com.cn/postDetail.do?view=1&id=2183080. Last retrieved on March 5, 2009.

34. bbs.people.com.cn/postDetail.do?view=1&id=84076662. Last retrieved on March 5, 2009.

35. bbs.people.com.cn/postDetail.do?view=1&id=1427828. Last retrieved on March 5, 2009.

36. bbs.people.com.cn/postDetail.do?view=1&id=83941623. Last retrieved on March 5, 2009.

37. Xinguang et al. (2003).

38. McCormick Freedom Museum, *America's Awareness of First Amendment Freedoms* (2006), www.freedommuseum.us/assets/pdf/e4/pressrelease/survey_results_report_final.pdf. Last retrieved on July 4, 2008.

IV

CONCLUSION

13

Online Chinese Nationalism(s)

Comparisons and Findings

Shaun Breslin and Simon Shen

THIS COLLECTION WAS CONCEIVED and put together as an attempt partly to bring the specific case studies to light, but also to use the analyses in these case studies to make a start at answering some bigger questions; not just to do with the role and significance of the Internet, but questions that relate to the changing basis of political power and the nature of the Chinese political system today. Perhaps most fundamentally, the question of whether the spread of the Internet provides a mechanism for individuals to not just participate in Chinese politics, but to influence the policy-making process resulting in actual change to actual policies.

Before going any further, it is important to point out that this study has limitations. As with all studies of the impact of the Internet on political processes (and not just in China), we recognize that Internet usage is not evenly spread across all sectors of the population. Although there are Internet users in the countryside, the average blogger or discussant is still most likely to be relatively young, relatively well-educated, and living in a city. The project also had some built-in limitations from the very beginning. Our intention was to provide a combination of detail and breadth; while this might sound like a contradiction in terms, the detail is provided by in-depth studies of individual postings on each topic, while the breadth is provided by considering as wide a geographic coverage as possible. In order to be able to handle the sheer weight of postings and also to provide a comparative basis for considering what we might call "the national temperature," the analyses have simply had to be limited by time and by outlet. In terms of temporal limitations, this volume contains all the caveats about the impossibility of being up-to-date that limit any book on contemporary politics (and not just Chinese politics).

In terms of outlet, the proliferation of blogs and the ability to leave comments on web pages means that there are so many Internet sites where opinions are posted that it is simply not possible to be comprehensive. There are also an increasing number of small specialized online communities that represent specific political viewpoints—as just one example, *wuyou zhixiang* has become a focal point for the promotion of Maoist viewpoints through a variety of online presences (Google group, blogs, bookstore website, etc). Whilst the postings on these specialists' sites are often extremely interesting, the attempt to find as far as possible what the national debates are on any given topic has meant that the majority of chapters in this collection have concentrated on the major national forums such as *Qiangguo*, Sina.com, and the *People's Daily* online (with some exceptions that we shall come to shortly).

In order to be as comprehensive as possible, whole regions have been covered in individual chapters. Such a regional focus in many ways reflects the way in which online debates in China also aggregate to the continental/subcontinental level; for example the tendency to often treat Africa as a single entity and to have a single idea and understanding of what "Africa" is and "Africans" are (which is far from just a Chinese phenomenon). Nevertheless, we accept that studies of individual cases within Latin America, Africa, and Europe might well generate different observations from considering regional states together—a task that we are at this point happy to leave to others to complete. So, with these caveats and pointers toward future research in mind, despite the different opinions and orientations of our authors, what can we glean from a collective reading of the contributions to this volume?

There are a number of conclusions that emerge from considering the contributions in this volume as a whole. By way of an "executive summary," these can be summarized under the following eleven themes:

1. Chinese politics is becoming more open and more pluralistic, and the Internet is playing a key role in facilitating this increased interest articulation.
2. But there is no strong evidence to date that this is leading toward Western-style democratization (as opposed to the democratization of one-party rule).
3. There is no single voice of Chinese online nationalism—rather the discussions are characterized by diversity.
4. Internet nationalism tends to be largely event-driven and responsive (though with some signs of a move toward agenda setting).

5. The main sources of Internet nationalism relate to external interference in issues of Chinese sovereignty and/or perceived external criticisms of China (the state and the people).
6. The United States and Japan are the main "culprits," and relations with the United States in particular are often the key determinant of how other countries/regions are perceived.
7. Internet nationalism is playing a role in the creation of new national/state identities.
8. Internet nationalism does appear to have an impact on policy-making and actual international relations.
9. Such nationalism at times runs counter to the foreign policy objectives of the leadership and creates problems.
10. There appears to be a trend toward mobilization and organization of ideas and opinions rather than just individuals articulating opinions.
11. The spread of Chinese Internet nationalism is reaching beyond the national scale and increasingly speaking to other public opinions/political processes.

The rest of this chapter puts some flesh on the outline skeleton of the main findings, starting with a consideration of the changing basis of one-party rule, and the role of the Internet in bringing about such change.

Pluralizing One-Party Rule

Although this book is concerned with Internet nationalism, the individual chapters have all been informed by wider debates about the impact of the Internet on Chinese politics per se and the prospects for democratization. And if we move beyond the specific focus of nationalism for a moment, the broader evidence points to an increasing pluralism of debate in contemporary China. This is not to say that all voices have equal weight, or indeed that all voices are listened to at all. There are clear limits over what can and cannot be said. Anything that challenges the right of the CCP to be the ruling party crosses the line and will be treated as illegitimate—for example, calling for really competitive multiparty democracy.

Along with these clear, well-established, and well understood limits—people know what is likely to happen to them if they call for the party to be challenged—there are also much less clear dividing lines over what is legitimate and what is not. Indeed, sometimes it is only clear that the line has been exceeded when the power of the state is utilized against offenders.

Furthermore, these blurred dividing lines also sometimes move, resulting in what was once considered to be permitted subsequently being reclassified as illegitimate. The lack of clarity emerges from distinguishing between what are illegitimate attacks on the system on one hand, and what are loyal criticisms from within an acceptance of the system on the other. Complaining about policy or even an individual official should be legitimate—but the blurring begins when such complaints are perceived as being critical of the system per se, and sometimes when the individuals concerned decide to silence their critics.

That said, the variety of what can be said and what is indeed said is much greater than before. For example, despite the considerable remaining sensitivities over June 4, 1989, a forum was held to consider its residual impact in Beijing in May 2009. And as is the practice now, the report of the forum on weblogs resulting in discussions about the consequences—including calling for support for the "Tiananmen Mothers" (of killed students) by holding a three-minute silence.[1] Yet we should not think that this is leading to attacks on party rule and the leadership—far from it. Indeed, an article on the above-mentioned *wuyou zhixiang* website ("Renren douying tiwen he huide de qida wenti: Dui sanshi nian gaige shida shiwu yuanyin he zeren de zhuiwen") generated wide-scale criticism by calling for Wen Jiabao to resign after finding a link to him for all of the "ten great mistakes" made in the thirty years of reform.

The ultimate removal of the piece criticizing Wen is perhaps not surprising (though it remains available elsewhere).[2] But even in these potentially critical stories and/or the responses to these stories, it is not the CCP's monopoly of power that is being questioned, but rather the failings within the overarching maintenance of party rule. Those who call for a reversal of verdicts on June 4 and justice for the victims are not asking for the party to stand down, but rather for the party to act justly. And as the Chinese leadership seeks to find new ways of legitimizing its rule, then the role of the general public in providing "surveillance" and/or checks and balances on the power of the party-state becomes not just legitimate, but actually encouraged.

For the leadership of Hu Jintao and Wen Jiabao, democracy in the Chinese context entails putting sovereignty back into the hands of the people. This does entail the extension of elections—but elections that pick and choose individuals at the local level, and do not challenge party rule per se. Indeed, the whole point about introducing new democratic reforms is to strengthen, not weaken, one-party rule. By showing that the people have the power to get rid of corrupt, weak, or simply unpopular local leaders, the hope is that party rule as a whole becomes more legitimate.

Democratization should also occur through popular surveillance of policy and officials on a daily basis. An administrative litigation law has been in place since as early as 1990, allowing individuals to use the law courts against local governments and officials to claim their rights and to protest abuses. More recently the Internet has been promoted as a key means of providing a link between the people and politicians—a mechanism through which people can complain, ask for help, and make suggestions. The focus is on individual action. But while the regime remains highly wary about anything that looks like coordinated activity, some NGOs have utilized the opportunities now available to them to lobby officials. Crucially, these groups act within the existing structure and don't wish to challenge it, typically simply asking for already existing legal rights to be properly and fully applied. For example, a group representing hepatitis C sufferers is very active in lobbying officials for the prevention of illegal discrimination against them (mostly in terms of employment).

Web sites have become almost totemic examples of the new "democratic" polity. Before the NPC session in February 2009, the image of Wen Jiabao at a computer terminal was on every TV screen and on the front of almost every paper on the newsstands in an attempt to show how responsive the leadership was to public opinion. Following a similar event the previous summer, Wen spent almost two hours answering questions submitted on www.people.com.cn, and virtually every newspaper, magazine, and media outlet set up special forums to discuss what was going on at the sessions and to post suggestions for possible topics for consideration.

Hong Yu, an NPC delegate from Wuhan and professor at Central Normal University, is credited with being the first delegate to establish an online presence designed to listen to popular views and to receive suggestions for policy changes in 2004—www.hongyu-online.com. There is now an evaluation system in Beijing and elsewhere whereby the general population is asked to comment on the work of local officials, with these evaluations published on official websites (which also become arenas for public comment). And as a final example here, the organization department of the CCP CC also established a website where people can even complain about the way that local officials are selected—www.12380.gov.cn.

So it is not so much that the articulation of interests through the Internet is tolerated, as it is being absolutely promoted by the state as a means of reengaging itself with the people (or, if you have a more cynical perspective, a means of making it look as if it is reengaging itself with the people). With the caveats relating to the limits of the acceptable in mind, encouraging popular participation in political debates via websites is considered to be an important means of reformulating the party-state structures of authority.

Establishing a New "State Identity"

Shared Historical Experiences, Benevolent Great Power Future

If establishing a more responsive party is one part of the attempt to shift the basis of CCP rule, creating a new state identity is a second key process. This new identity revolves around the idea that China is "different"—what we might call "Chinese exceptionalism." And this promotion of Chinese difference is not only a cause of Internet nationalism, but is also supported by it as well; there is a feedback loop between official positions consciously established by China's elites, and popular manifestations of national sentiments as shown in online discussions through which each reinforces the other.

In the international realm, there is an attempt to create an understanding of China as a different sort of great power—different from the way that previous great powers have behaved in the past and different than how the West (which really usually just means the United States) acts as a hegemonic power in a unipolar world today. This partly entails a rejection of and revulsion against such hegemonic practices—be these the way that foreign powers unfairly subjugated China through force in the nineteenth century, or the way that the United States exercises its hegemonic power today. In the case of the United Kingdom, Lee notes that these historical and contemporary actions are combined by linking UK involvement in Iraq with what it previously did in China. And in general, the extent to which countries ally themselves with the U.S.-led promotion of supposed universal liberal political and economic norms is an important indicator of how these countries are perceived and depicted in online Chinese communities (an issue we will return to shortly).

But this discourse is not just about how China perceives others; how China perceives itself and its own conduct of international relations is equally if not more important (and perhaps increasingly so).[3] To be sure, this self identification is largely specifically established as a direct response to the actions and philosophies of Western powers. For example, Shen's chapter on Africa, Zhang's on Southeast Asia, and, to a lesser extent, Leung's on Latin America all point to the perceived significance of historic friendly relations as a basis for contemporary interactions—among them, a shared history of colonization and/or subordination to arrogant and powerful Western powers. And, increasingly, understandings of the way that China acts—and will act in the future as it becomes ever more powerful—are framed in terms of Chinese "difference."

So, as Liu points out, when the United States/the West is being criticized for the promotion of U.S./Western interests and values, this partly reflects a real resentment at the perceived inequity in the global system. But it also partly represents a resentment that China is not able to punch its real weight in the international system because of the way that existing structures unfairly empower Western states at the expense of emerging developing states. Moreover, it also partly provides the "other" against which putative Chinese great power action (and identity) is constructed. Thus, in online discourses toward Africa and Southeast Asia, we see the transition from conceptions of partnership and equality toward Chinese superiority and dominance. Crucially, this newly superior power is perceived as being very different from previous great powers—China does not impose its values on others, but instead respects the sovereignty of others, and is a benevolent and responsible great power that promotes a democratic global order which will help generate security and prosperity for all.

The "Other" as Alternative to China's "Success"

For Shen, the online community fulfills another role in helping to reinforce a new state identity in that China's economic success is contrasted with perceived continued poverty in Africa. The same message comes across in Leung's analysis of perceptions of Latin America. And again here we see a feedback loop where perceptions of the region are shaped by domestic debates and discourses, and the resulting perceptions of the region reinforce the emerging state identity which shapes the domestic discourses. Here, conceptions of development in Latin America are shaped by the domestic discourse in China about the dangers of what is known as Latin Americanization—a situation of massive inequality exacerbated by endemic corruption.

As such, online discussions typically conceive Latin America as "backward" and corrupt in contrast to China's own economic success and dynamism (though we should note that a fair amount of the domestic online political discourse focuses on the apparently endemic state of corruption in China itself). However, this general understanding is tempered by the possibility of a socialist revival in the region which generally generates positive comments—and interestingly, each county's relationship with the United States is linked to perceptions of that country's development. At the risk of oversimplification, a close link with the United States means increased social stratification, massive inequality, and social strife. Anti-Americanism equates with socialist revival and the potential to emulate China's own success.

The Other and Chinese Uniqueness

So what we have here is not just an emerging national pride at China's economic success, but pride that China has done as well as it has in its own way; it has eschewed the dominant prescriptions of "the Washington Consensus" and developed its own indigenous path to great (economic) power status. There are some signs that Chinese thinking is beginning to change. Not least through the popularization of Ramo's "Beijing Consensus"[4] in China (known in Chinese as *Beijing Gongshi*), the idea of a Chinese alternative for others to follow is gaining some purchase.[5] But by and large, the emphasis remains on the uniqueness of China. In the words of Pan Wei:

> [The] China model consists of four sub-systems, they are: a *unique* way of social organization, a *unique* way of developing its economy, a *unique* way of government, and a *unique* outlook on the world.[6] (emphasis added)

The "profound lesion" lesson that China provides for other developing states is that each country must do what suits it best based on the reality of its own special and unique circumstances: "start from national conditions, take your own road"—*cong guoqing chufa, zou ziji de lu.*[7]

The roots of this difference are found in China's history and in an eclectic use of a variety of thinkers and philosophies; of Confucius and Mencius, of Lao Zi and Zhuang Zi, of Sun Zi, and *The Art of War*. China is where it is today because of where it was. As such, trying to impose external supposedly universal ideas, norms, and expectations on China is at best misguided—they are not universal but reflect the specific and unique historical evolution of other (Western) societies and so simply cannot be applicable to China (or indeed to anywhere else). At worst, any attempt to impose external values under the guise of universalism represents a deliberate attempt to prevent China from rising and assuming its rightful place in the world.

Moreover, in some arenas at least, China's way of doing things appears as not just best suited to China's own circumstances, but superior to dominant Western norms. Take, for example, this excerpt from the White Paper on the Rule of Law issued in February 2008:

> China has a 5,000-year history of civilization. And the Chinese legal system goes back to ancient times. As early as in the 21st century BC, consuetudinary law appeared in China's slave society. In the Spring and Autumn and Warring States periods (770–221 BC), written law was promulgated in China, and a systematic written code of laws appeared. In the Tang Dynasty (618–907), China had a fairly complete code of feudal laws, which was passed on and developed in the following feudal dynasties. The Chinese system of law emerged as a unique one in the world.

What is explicit here is that China's legal system has emerged over thousands of years to suit China's circumstances. What is implicit is the superiority of a legal system that actually predates Western forms of legalism by centuries. The focus on harmony, collectivism, peace, and virtue as the basis of Chinese society also implicitly (and occasionally explicitly) sits in stark contrast to the Western tendency to use force to immorally impose its own values that value individualism and greed.

Sources of Online Nationalism

The State, The Leaders, and The People

This suggests, then, that there is a latent distrust of the West and an emerging latent feeling of Chinese exceptionalism—if not yet Chinese superiority. We can also add to this suspicion (at the very least) of Japan. Yet in "normal" times, online expressions of nationalism tend to be rather muted. These latent concerns occasionally result in critical postings and comments, but for the most part day-to-day discussions of the rest of the world are not emotive or necessarily politicized. As noted in the chapters in this collection, discussions are often dominated by talk about scenery, holiday destinations, football matches, education systems, how to do business, and so on.

To be sure, these narratives are often imbued with nationalist sentiments representing the identity formation we have discussed above. Shen's discussion of the stereotyping of Africans is a good case in point. But for the most part, Internet nationalism in its most vociferous form is event-driven—something happens that rallies the online population to action. Given the time period during which the contributors to this volume were undertaking their research, the two examples of the Olympic torch relay and the Tibetan disturbances in 2008 have reoccurred throughout this book. The Tibetan example is particularly important here as popular uproar in China was caused by foreign responses to the event rather than the event itself (as was the case with the torch relay).

To try to understand the extent of popular feeling, we need to return to the start of this book and Tok's discussion of the importance of sovereignty, and add to it Alistair Johnston's argument that the distinction between the Chinese state, China's leaders, and the Chinese people is very weak—or, put the other way round, there is a strong and deliberately constructed link between the three.[8] So if the policies pursued by China's leaders are criticized, then this is taken as a criticism of not just the leaders, but "China" as a whole. So when demonstrators express support for the Dalai Lama, this is taken as an attack not just on the Chinese state, or of the policy of China's leaders, but as an attack on China as a whole including the Chinese people. Thus, as Pan notes in his chapter on perceptions of the United States, when U.S. journalists claimed that their negative remarks about Chinese actions in Tibet were aimed at the government and not the Chinese people, this was simply dismissed as sophistry in China. Indeed, Pan's own comment on the affair in this collection reveals the apparently inextricable link between the people and the state—"CNN's attempt of separating Chinese government and people is so naive that it deserves to be boycotted."

This association of people, leaders, and state is manifest in those official foreign ministry statements that point to how the actions of others have hurt the feelings of the Chinese people (*shanghai zhongguo renmin ganqing*). One Chinese blogger went back through the archives of the *People's Daily* online and discovered that the Chinese people's feelings had been hurt a total of 115 times between 1946 and 2006 by fifteen countries, the Vatican City, the European Union, NATO, and the Nobel Prize Committee.[9] The single most offending country was—perhaps not surprisingly—Japan, with forty-seven cases, followed (again not surprisingly) by the United States with twenty-three. These slights have included criticisms of China's paramount leaders—Albanian attacks on Mao Zedong and the Japanese foreign ministry's criticism of Deng Xiaoping. But the overwhelming majority of the rest involve foreigners interfering in some way with matters pertaining to Chinese sovereignty. These include a real physical attack on sovereign territory in the shape of the NATO bombing of the Chinese embassy in Belgrade, border disputes with India, and the higher authority of the Vatican over Chinese Catholics.[10] But by far the two most reoccurring sources of hurt feelings are acts that relate to the possible independence of Taiwan and support for the Dalai Lama, who is accused of attempting to split China and establish a separate Tibetan state.

The importance of Taiwan and Tibet highlight one of the key themes in Chinese nationalist discourses. As with nationalisms in other countries, what others do or have done in the past or might do to harm the perceived national interest are of course important. But perhaps more so than with any other country, so too is this key theme of not interfering with issues that are wholly domestic in nature and therefore not legitimate areas for external comment. Crucially, this is more than just not liking what others say about a domestic issue. In virtually every country in the world, there will be a nationalistic response of some sort to external criticisms of domestic politics, society, and culture. But in the Chinese case, it is not just that the criticisms and comments are harmful or hurtful, but that they are made at all.

As Wong shows in his chapter on Taiwan, the very process of commenting on something that is conceived of/depicted as an entirely domestic affair falling wholly within China's sovereign affairs is in itself considered illegitimate—the making of the comments are in themselves an attack on China that should be resisted.[11] Establishing a division between making the criticism in the first place and the content of the criticisms itself might seem like an exercise in splitting hairs—they both result in "hurt feelings" and arouse nationalist responses. But there is an important difference in terms of the responses that the criticisms engender—the latter requiring a refutation of the argument, with the former simply requiring the argument to be rejected as invalid without any further reason or explanation. To be sure, Wong points out that much of the nationalist rhetoric relating to Taiwan revolves around the idea that major states—notably Japan and the United States—actually support Taiwanese independence (despite official policies), but the bottom line that the unification of China is not a legitimate sphere for external comments and criticisms remains.

Targets of Nationalism

Important as Taiwan is, Leung identifies surprisingly little online interest in Latin America, even though this is one of the parts of the world where diplomatic rivalry between China and Taiwan remains important, with Taiwan still having considerable support. Indeed, it's probably the only part of the world where Taiwan has a major formal diplomatic presence. Although Dominica switched recognition to the PRC in 2004, Grenada in 2005, and Costa Rica in 2007, St. Lucia switched back to Taiwan from the PRC in 2007 (having previously switched from Taiwan to the PRC in 1997), and at the time of writing twelve of the twenty-three "states" that recognize Taiwan are in Latin America and the Caribbean.[12] Rather, as we have already seen, how Latin American countries are perceived is in large part shaped by their relationship as ally or opponent of the United States. Similarly, we have already noted how attacks on the United Kingdom were inspired by slavishly following the United States into war in Iraq.

In Southeast Asia, Zhang identifies Taiwanese attempts to develop a quasi-diplomatic presence and the roles that the United States and Japan play in the region as the major drivers of online debates. And the United States and Japan remain the main focus of online nationalism. Indeed, if we count the ten times that NATO hurt the Chinese people's feelings as at least partly an attack by the United States, then Washington and Tokyo combined account for almost 70 percent of the insults. Japan not only has its own online discussion forum—the China-Japan Forum—but is also the target of some of the most passionate postings, which at times verge on xenophobia.

There is also something of a tendency to associate the comments of individuals and/or media outlets with countries as a whole—the reverse side of the above-mentioned correlation of people, leaders, and state in China. This also reflects an externalization of Chinese state-society relations onto understandings of how other countries work. For example, understandings of the way that the media functions within China as some form of agent of official policy seems sometimes to be transferred onto the role of the media in other countries. The fact that newspapers in the United Kingdom or the United States spend most of their time being critical of their own national governments is often overlooked, and their criticisms of China sometimes at least are taken as being representative of that country's attitude towards China as a whole (if not necessarily equated with official policy).

Indeed, in the European case, there was an extra degree of blurring between nonstate, state, and super-state, as the actions of people in France not only impacted on Sino-French relations, but China's relations with the European Union, as France held the EU presidency at the time. Or, conversely, the protest against what happened in France was first given a national identity—it was a French issue rather than just by some French—and then materialized in political action against companies that were also identified as being part of this thing called France, including Carrefour, which, as King points out, has largely localized its activities in China and primarily sells Chinese-produced goods!

We should not push this too far. Pan notes that the Chinese online community reserved specific criticism and opposition to CNN for negative comments about China rather than turning the focus on the United States as a whole. And there is an increasing recognition of diversity of perceptions about China overseas. Nevertheless, the possibility for actions of individuals and groups to be taken as representative of some higher form of national effort remains—as does the likelihood that companies may be punished for the actions of governments, groups, and individuals beyond their control.

Popular Participation and the Policy Process

So the changing basis of legitimacy for single party rule in China impacts on online debates (and is reinforced by online communities) in two main ways. First, it entails the creation of a more pluralistic, listening, responsive, and accountable system of rule. Second, it entails the establishment of the idea of China as different, and of China's harmonious and virtuous values standing in contrast to the individualism and confrontational values of the West—values that the West arrogantly tries to impose on others, a process that the CCP resists on behalf of the Chinese people.

In combination, these two factors create a situation that legitimates and indeed encourages popular online participation in discussions that promote patriotism and protest against external slights and attacks on China. As Liu argues, the state has attempted to "co-opt" the public and get them to share the state's objectives through encouraging "'rational' and 'constructive' patriotism." And notwithstanding the increasing provision of legitimate online forums that allow individuals to take part in political discussions and to try and influence the political process, patriotism/nationalism remains one of the "safest" issues for the general public to engage with. Nationalism has become not so much *a* medium for popular participation as *the* key vehicle through which individuals can participate and articulate their interests and identities in the Chinese political sphere.

Clamping down on popular nationalism runs the risk of not only making the party look out of step with patriotic public opinion, but also of removing the one political sphere that has emerged as a legitimate medium for interest articulation and thus could decrease rather than reinforce regime legitimacy.

Diverse Online Voices

In considering the way in which online debate influences policy, it is important to avoid the trap of thinking in terms of a single Chinese nationalist voice. Indeed, ironically, one of the key conclusions that we can draw from the chapters in this collection is that it is not easy to come to simple overarching conclusions. Instead, what we see is an increasingly diverse set of interests and demands which make it extremely difficult for the government to respond to them all. In some cases, the extremes of interests articulated make it impossible for any response to satisfy all concerned.

Perhaps the clearest example of the diversity of different approaches comes from Pan's analysis of perceptions of the United States. Here we see positions that change from outright hatred to a relatively strong strand of opinion that expresses great admiration for the U.S. society, the U.S. way of life, and the idea that the United States is still the land of opportunity where small and ordinary people can and do achieve their dreams. But such varying attitudes, voices, and interests are found across the range of countries and areas discussed in the chapters in this book. Even when it comes to Japan, while there is not much admiration present in online discussions, there are varying interpretations of how China should respond and the demands that online communities make for action.

In general, it appears that the more knowledge that online communities have of a country or a region, then the more diverse and more balanced opinions tend to be. This is perhaps most clearly the case in relation to Southeast Asia, but there also appears to be a growing depth of knowledge of the United States, which has generated the above-mentioned admiration for some parts of U.S. society which sits alongside approbation for U.S. foreign policy. Conversely, a lack of knowledge tends to result in stereotyping which in turn results in a much more homogenous online position—for example, perceptions of Africa and Latin America (a position that is far from unique to China).

So to say that the Chinese government is responding to online communities is correct to an extent, but can be misleading if it is built on an understanding of a homogenous Internet community. Rather than respond to *a* voice, the task is the more complex one of deciding *which* voice or voices to listen to, and how to balance contending demands. Crucially, Reilly shows how popular discussions and debates politicized the debate over the possibility of adopting a Japanese HST system, moving the decision-making system out of the hands of a small bureaucratic group. What had previously been largely a technical issue relating to cost and feasibility was politicized by Internet activism, and brought into the realm of high politics and national strategy. And ultimately popular pressure ideologicized the debate and meant that pragmatic practical, technical, and economic considerations were trumped by political concerns.

Responding to Online Debate

The extent to which online debates reinforce the stereotypical image China has of the outside world remains an issue to be studied in more depth.[13] There is, however, prima facie evidence from the studies in this collection to suggest that policy-making elites in China are indeed adopting ideas first aired by Internet users to formulate Chinese foreign policy. The HST is a good example here. Similarly, Chinese scholar Zhou Mi argues that the sudden cancellation of her scheduled meeting with Japanese Prime Minister Koizumi, as well as the schedules of other high-level meetings, can be viewed as being influenced by online protests about China's foreign policy.[14]

The extent of popular hostility toward Japan perhaps makes it something of a special case, and demands for a more aggressive Chinese position vis-à-vis other countries and regions appear to be less common. But they can be found, and as King notes (referring to Pei Minxin) in the European context, radical "fringe elements" are often louder than their representation in society as a whole deserves. Similarly, Zhang points to the numerically small but relatively vociferous group of netizens who call for a more forceful presence and policy in China's "backyard" in Southeast Asia. Pan might be correct to say that foreigners sometimes pick these extreme views to reinforce their existing negative perceptions of China, but that does not mean that these minority views don't exist at all. Indeed, when it comes to Sino-Japanese relations, extreme views on either side often result in nationalist responses in the other which in turn generate their own nationalist reaction. As such, this spiral effect has the potential to pull extreme positions into the mainstream.

Organization and Articulation

As in all societies, then, the intensity of an emotion can result in it being articulated in a manner that overrides more widely held but less intensely felt beliefs, particularly if there is a concerted mobilization of the like-minded to push the message home. And such mobilization does indeed seem to be an increasing element of online interest articulation.

It is going too far to suggest that the Internet provides a sphere of public activity that is out of the control of national governments. Internet sites and ISPs are subject to official surveillance and monitoring, and some have been closed by the authorities. Nevertheless, the sheer number of users and potential outlets and the speed at which postings are made in response to news stories means that it is extraordinarily difficult to have total control all of the time—and for Benson Wong, much harder to control than even the massive proliferation of mainstream print media outlets.

Very much aware of the regime's almost paranoid dread of anything that looks like organized political activity, the individual nature of Internet postings has allowed individuals to participate in legitimate ways. Building on Reilly's analysis, we can argue that as the Internet is becoming an ever bigger source of information and news for many, online campaigns develop their own life-cycles as people (particularly the young middle classes) post their opinions in chat rooms and on discussion groups. Thus, a movement of sorts emerges as a result of the aggregation of myriad individual actions.

But there also seems to be something more coordinated emerging. As Peter Gries and Shih-Diing Liu respectively argue, when a series of anti-American and anti-Japanese demonstrations broke out in China in 1999 and 2005 respectively, online mobilization was the major catalyst.[15] They claim that this online mode of mobilization bypasses the censorship of the party-state in China, and represents a bottom-up approach of popular nationalism (*renmin minzuzhuyi*). For Liu in this volume, they are "grassroots organized campaigns" engaged in "collective action" to lobby for policy change.

Nationalism as "Problem"

Perhaps not surprisingly, extreme positions are often picked up overseas and used as evidence to point to an increasingly nationalistic and xenophobic China. And this brings us to the extent to which online nationalism is a help or a hindrance to China's state elites. So far, we have emphasized the extent to which nationalism reinforces the regime. In the post-Mao era—and perhaps particularly since Tiananmen in 1989—the leadership has attempted to portray itself as the defenders of China's national interests in a hostile international environment, with major powers committed to preventing China's rise. The inculcation of these sentiments within the general public establishes a common purpose between party and people and relegitimates the CCP's monopoly on political power.

But there is also an awareness that this is a potentially dangerous game to play—the promotion of nationalism might create expectations that the party cannot deliver on; or domestic nationalist aspirations might jeopardize foreign policy objectives; or worse still, both in combination. As already noted, it is difficult for the leadership to clamp down on nationalism without damaging its self-identified role as vehicle for the articulation of national interests.

Perhaps the key here is the distinction between the national identity that China's leaders want to establish for domestic legitimation purposes, and the promotion of China as responsible great power (*fu zeren de daguo*) for an international audience. Thus, Benson Wong argues that the activities of anti-Taiwan nationalists in China might undermine the government's attempts to depict China as peaceful and benign and a force for peace and stability. Qiu warns that some radical users might hack into what they see as anti-Chinese websites, similarly undermining China's international image overseas.

We have already seen examples of how these domestic and international impulses have conflicted with each other in the ultimate drawing back from the overzealous promotion of a rather radical nationalist position against Japan. King similarly argues that the nationalist response to problems with the torch relay in Europe destabilized Sino-EU relations—and couldn't have come at a worse time in the development of the relationship.

Chinese Online Nationalism Goes Global

Crucially for King, the European case entailed an additional level of participation in that it entailed coordination and mobilization by Chinese communities across the world on online platforms that are even less controllable than those that fall within Chinese jurisdiction. The extent to which this indignation spilled out from Chinese communities across the world was indeed significant. The speed in which Chinese counter-protestors were mobilized to support the torch relay and oppose the pro-Tibet protestors is testament primarily to the strength of feeling amongst Chinese across the world—but to not give credit to the organizational abilities of the Chinese authorities within Chinese (largely student) communities overseas would be to remove the state too far from the equation. This went much further than just the very visible image of Chinese flags being waved by supporters in San Francisco and elsewhere. Chinese voices were very prominent on non-Chinese language discussion forums and in posting comments on news stories across the world.

Whilst these responses in part simply represented the articulation of anger and frustration, there was also a message in many of them to local populations. This was an effort in part at least to influence public opinion and political debates within the national sphere of the postings—in the United States, France, Britain, and other countries where there was perceived "China bashing." And, indeed, the combination of global outrage and a domestic boycott of French producers and goods in China resulted in the French government and French companies responding with initiatives designed to win back support in China.

Indeed, while the focus on this collection has been explicitly on Chinese Internet nationalism as manifest in Chinese language contributions, King's study of the European Union highlights the importance of Chinese overseas participating in foreign language Internet debates designed to influence debates about (and policies toward) China in the rest of the world. As Lee notes, Chinese students' online activism in the United Kingdom seems to be playing an increasingly important role in shaping Chinese views of the United Kingdom per se through their communication with friends and ex-classmates back home.

For Liu, the activity of this transnational Chinese community in 2008 shows that Internet activity and activism is threatening Westphalian territorial conceptions of loci of political activity and even the nature of global politics itself. And the evidence in this collection points to ways in which the Internet results in the increasing porosity of political borders, and globalizes arenas of political activity and interest articulation. You do not have to be a national citizen and a registered voter to post opinions or mobilize behind a cause. You don't even need to have visited the country where you are posting your views. But this does not mean that political borders are wholly irrelevant, as the domestic political context continues to influence both what is said and how it is said.

The majority of the discussion in this volume explains the way in which the specific context of Chinese politics shapes the way in which Internet nationalism can be articulated and organized and the constraints on such activity. Despite the transnational nature of Internet activity, it is fair to suggest that activity outside China will also be shaped by the nature of the political system and public opinion that it is trying to shape. Indeed, King shows how Chinese participation in debates over China in the United Kingdom and France were self-consciously intended to correct what the activists thought were deliberately unfair and misinformed understandings of China and Chinese politics. What we see is Chinese nationalist voices not only increasing in volume within Chinese politics, but also within the politics of other countries—deliberately and explicitly constructed to influence opinion and policy within those other countries. Thus transnational nationalism remains conditioned to some extent by the national political system in which it is located (even when it goes overseas).

Placing Shen's consideration of anti-Japanese nationalism in Hong Kong alongside the other chapters that focus on action on the mainland provides a clear example of how the national setting in part shapes the nature of the debate—the content of the debate, how it is articulated, and its target in terms of influencing political actors and engendering political change. To be sure, Hong Kong is an exceptional case, as it is a Chinese territory. As such, and even though Beijing is responsible for the diplomatic affairs of the territory, there is an alternative government to lobby to protect Chinese interests when it comes to Hong Kong's Chinese nationalists. And if there is wider frustration with the government, then the use of nationalism as a vehicle for expressing dissatisfaction is deployed against a local in the shape of the Hong Kong authorities, rather than against Beijing. But the clear, and at times self-defined, differences between different "types" of Chinese Internet actors in Hong Kong and the mainland is particularly noteworthy.

Final Words

With the study of Internet nationalism in China in its very early stages, there remains huge potential for the extension of research in further studies. We have already pointed to some of these in this conclusion: a focus on specialist groups/sites, more individual country studies in different regions, and the significance of Chinese postings on non-Chinese language media across the world. In addition to this, we suggest that disaggregating Chinese netizens would also be an interesting but extraordinarily difficult task—for example, filtering for age, sex, social status, relationship to the party, geographic location, experience of traveling abroad, and so on. A greater understanding of the mechanics of foreign policy-making and how public opinion via the Internet feeds into these processes would also be illuminating, but equally difficult to achieve.

Despite its relative infancy, we have already seen Internet nationalism play important roles in shaping not just wider public opinion, but also government policy toward Japan and the European Union/France. In the second of these examples, online communities played important roles in coordinating activities that also resulted in policy change in France itself in response to the strength of Chinese feeling. Evidence suggests an increasing confidence in the successes, longevity, and even possibly transferability of the Chinese economic model. China is also playing an increasing (and increasingly visible) role in the global economy, and in global economic governance. There is a general demand for China to be given the recognition that it deserves—recognition in terms of formal participation and (voting) power in global forums, and recognition in a less tangible form of "respect"—respect for China's achievements and respect for China's different understanding of political, economic, and social norms. In this situation, the likelihood that the actions of at least some foreigners will be deemed to be lacking respect and hurting the feelings of the Chinese people must be very high. Whether the government can manage to show its own nationalist credentials by responding to popular demands while simultaneously portraying an outward image of international responsibility remains to be seen.

Notes

1. Bullogger.com, www.bullogger.com/blogs/mozhixu/archives/295142.aspx, accessed 22 May 2009.
2. And was no longer available on wyzx when we tried to access it in May 2009. It was, however, reposted on www.reviewing.cn/article/2008/1005/article_2848.html, accessed 22 May 2009.
3. Of course, "China" here is used to refer to the collective words and ideas of Chinese people, and does not suggest that states/countries have identities independent of those constructed by human beings.
4. Ramo Joshua, *The Beijing Consensus Notes on the New Physics of Chinese Power* (London: Foreign Policy Centre, 2004).
5. Young Nam Cho and Jong Ho Jeong, "China's Soft Power: Discussions, Resources, and Prospects," *Asian Survey* 48:3, 2008, 264.
6. Pan Wei, "The Chinese Model of Development," Presentation at the Foreign Policy Centre, 11 October 2007, available at fpc.org.uk/fsblob/888.pdf, 2, accessed 30 January 2009.
7. Shen Li and Bai Qunying, "Jiedu Zhongguo Jingji Moshi (Analysis of China's Economic Model)," *Guangming Ribao*, 15 May 2006.
8. Alistair Iain Johnston, *Social States: China in International Institutions, 1980–2000* (Princeton, N.J., and Oxford: Princeton University Press, 2008).
9. Fang Kecheng's blog, posted 9 December 2008 at www.fangkc.cn/thought/donot-hurt-chinese. We were alerted to this blog by a comment on it in Joel Martinsen, "Mapping the Hurt Feeling of the Chinese," posted 11 December 2008 on www.danwei.org/foreign_affairs/a_map_of_hurt_feelings.php.
10. Specifically, the hurt derived from the Vatican's canonization of Chinese saints (fengsheng), but this is part of a wider dispute over who has the right to have the supreme authority over Catholicism in China—the Vatican or the Chinese state.
11. In theory, of course, this means that external comments that agree with the Chinese position on Taiwan and Tibet should also be illegitimate as they are also getting involved with a sovereign/domestic issue—but of course the *China Daily* in particular seems to take delight in publishing overseas opinions that concur with dominant Chinese views.
12. "States" in quote marks here, as one of these is the Vatican.

13. Mark Warschauer, "Language, Identity, and the Internet," in B. Kolko, L. Nakamura, and G. Rodman, eds., *Race in Cyberspace* (New York: Routledge, 2000), 151–70.

14. Zhou Mi, "The 2005 Anti-Japanese Demonstration in China: the Rise of Popular Nationalism," in Chinese Papers Collection Online of University Service Centre for China Studies, Hong Kong: University Service Centre for China Studies, 2007 [www.usc.cuhk.edu.hk/wk_wzdetails.asp?id=6212, accessed 30 January 2009].

15. Peter Gries, "Tears of Rage: Chinese Nationalist Reactions to the Belgrade Embassy Bombing," *The China Journal*, No.46, July 2001, 25–43; Shih-Diing Liu, "China's Popular Nationalism on the Internet. Report on the 2005 Anti-Japan Network Struggles," *Inter-Asia Cultural Studies*, 7:1, 2006, 144–55.

Index

Note to index: An *f* following a page number denotes a figure on that page; an *n* following a page number denotes a note on that page; a *t* following a page number denotes a table on that page.

About the Editors and Contributors

Shaun Breslin is a professor of politics and international studies at the University of Warwick. His latest book is *China and the Global Political Economy* (2007), and he has previously authored books on Mao Zedong, center-local politics in China, and, with Rod Hague and Martin Harrop, comparative politics. He has published dozens of academic articles in the previous decades, and is also the coeditor of *Pacific Review.* In addition to his interest in the (international) political economy of China, he also writes on theories of comparative regional integration.

Winnie King is a research fellow at the University of Bristol Centre for East Asian Studies and a member of the British Interuniversity China Centre, United Kingdom. Dr. King specializes in the political economy of greater China, East Asian regionalism, and political mobilization processes and policy-making in the political and economic development of Asian countries. She has written on Taiwan's socioeconomic relationship with China in *Chinese History and Society* (2007), and is currently writing a book on policy-making and political mobilization in an East Asian context. She is also editing a book in Sino-European relations (forthcoming).

Chun-wing Lee obtained his PhD in sociology at the University of Manchester. He is currently a teaching fellow in the Department of Social Sciences at the Hong Kong Institute of Education. His research interests include nationalism, class analysis, social movements, and the political/sociological aspects of sports.

Kai-chi Leung received his PhD in geography at the University of Minnesota. His research interests includes urban studies, technology and society, Chinese diaspora, globalization, and international development. He is currently a part-time lecturer at the Department of Geography, University of Hong Kong, and the Department of Cultural Studies, Lingnan University. Besides his academic commitment, he also shows strong interest in community participation, including various local urban contestations. He often serves as a guest host of a radio program on Hong Kong commercial radio, and is a regular contributor to *Ming Pao Daily* and *CUP* magazine.

Shih-Diing Liu teaches in the Department of Communication, University of Macau. His recent publications appear in *New Left Review, Positions: East Asia Culture Critique, Reflexion*, and *Taiwan: A Radical Quarterly in Social Studies.* His current research focuses on the theoretical implication of Internet politics for China's postsocialist hegemonic formation.

Yaling Pan is the assistant dean of the Institute of International Studies, and associate professor of the Center for American Studies, Fudan University. She was granted a bachelor's degree in 1998, master's degree in 2001, and doctoral degree in 2007, all on international relations. She has published a book titled *American Patriotism and U.S. Foreign Policy* in 2008, more than thirty academic papers in Chinese and English journals, and many articles in newspapers. Her major research interests include American political culture, Sino-U.S. relations, and international relations theory. Currently, she is writing a book on migrant's patriotism, lobbying, and U.S. foreign policy, and doing research on U.S. foreign policy during the Cold War.

James Reilly is a lecturer in the Department of Government and International Relations at the University of Sydney. His research and teaching are in the areas of Chinese foreign policy, East Asian politics, and international relations. He is currently preparing a book on the rise of public opinion in Chinese foreign policy. He is the author or coauthor of articles in *Survival, Washington Quarterly, Asian Survey*, and *China: An International Journal.* He holds a PhD in political science (George Washington University, 2008) and an MA in East Asian area studies (University of Washington, 1999). Before joining the University of Sydney in 2009, he was a postdoctoral research associate with the China's War with Japan Programme at the University of Oxford.

Simon Shen is an associate professor in the Department of Social Sciences, Hong Kong Institute of Education, and an adjunct associate professor and coordinator of the Master of Global Political Economy Programme, Chinese University of Hong Kong. Trained as an international relations scholar, he got his DPhil from Oxford University and his MA/BA from Yale University. Previously he has worked in Tsinghua University, the Hong Kong University of Science and Technology, and the Chinese University of Hong Kong. His current researches include Chinese nationalism and Chinese foreign policy, external relations of Hong Kong, anti-terrorism, and regional security in Asia. He is the author of *Redefining Nationalism in Modern China: Sino-American Relations and the Emergence of Chinese Public Opinions in the 21st Century* (2007), and has published numerous articles in leading journals such as *Asian Survey*, *Pacific Affairs*, and *Pacific Review*.

Sow Keat Tok is doing his PhD in the Department of Politics and International Studies, University of Warwick. His dissertation examines China's evolving idea of the "state," and its corresponding "sovereignty" in the light of central-local relations, by taking a comparative approach and using the case studies of Hong Kong and Taiwan. He is also the current English editor for the Chinese periodical *Journal of Contemporary Asia-Pacific* [*Dangdai yatai*]. Previously, he has held research positions in the China Policy Institute, University of Nottingham, and, prior to that, in the East Asian Institute, National University of Singapore. His research interests cover China's leadership politics, central-local relations, foreign relations, and foreign policy thinking.

Benson Wai-kwok Wong is assistant professor in the Department of Educational Policy and Administration at the Hong Kong Institute of Education. He is serving as a coinvestigator of the Hong Kong Transition Project. Previously he was a fellow in the Center for Contemporary China, National Tsinghua University, Hsinchu, and a postdoctoral fellow in the Taiwan Foundation for Democracy, Taipei, Taiwan. His research has focused on education (teachers) and politics, media (Internet), political interactions in greater China, and political symbolism. He is presently working on a project about China study tours for local students and the reconstruction of the China/Hong Kong identity. He has published articles on China's Hong Kong policy in the pre-1997 era, iconographic study of Hong Kong's and Taiwan's electoral campaigns, as well as political posters in mainland China, and political participation of teachers as well as teacher unionism in Hong Kong and Taiwan.

Chun Zhang is a research fellow of the Institute for Foreign Policy Studies, Shanghai Institutes for International Studies (SIIS). He is also taking charge of the academic management of SIIS. His researches focus on Chinese Africa policy, Northeast Asia studies, and international relations theory. He gained his doctoral degree from Fudan University in 2006. He has published several books, with the *American Think Tank's Influences on "One China" Policy* (2007) rewarded with a First Class Award (the highest level) of the Ninth Excellent Philosophy and Social Science Publications Award of Shanghai in October 2008. He has participated in three books' translation (from English to Chinese), and published more than thirty academic papers.

www.ingramcontent.com/pod-product-compliance
Lightning Source LLC
Chambersburg PA
CBHW031415220926
51612CB00004B/36
* 9 7 8 0 7 3 9 1 3 2 4 7 0 *